Algorithms and Complexity

Herbert S. Wilf

Professor of Mathematics
University of Pennsylvania

Prentice-Hall, Inc., Englewood Cliffs, New Jersey 07632

© 1986 by Prentice-Hall, Inc.
A division of Simon & Schuster
Englewood Cliffs, New Jersey 07632

Printed in the United States of America

10 9 8 7 6 5 4 3 2

ISBN 0-13-021973-8 025

Prentice-Hall International (UK) Limited, *London*
Prentice-Hall of Australia Pty. Limited, *Sydney*
Prentice-Hall Canada Inc., *Toronto*
Prentice-Hall Hispanoamericana, S.A., *Mexico*
Prentice-Hall of India Private Limited, *New Delhi*
Prentice-Hall of Japan, Inc., *Tokyo*
Prentice-Hall of Southeast Asia Pte. Ltd., *Singapore*
Editora Prentice-Hall do Brasil, Ltda., *Rio de Janeiro*
Whitehall Books Limited, *Wellington, New Zealand*

Contents

Chapter 0: What This Book Is About

Chapter 1: Mathematical Preliminaries

Chapter 2: Recursive Algorithms

Chapter 3: The Network Flow Problem

Chapter 4: Algorithms in the Theory of Numbers

Chapter 5: NP-completeness

Preface

For the past several years mathematics majors in the computing track at the University of Pennsylvania have taken a course in continuous algorithms (numerical analysis) in the junior year, and in discrete algorithms in the senior year. This book has grown out of the senior course as I have been teaching it recently. It has also been tried out on a large class of computer science and mathematics majors, including seniors and graduate students, with good results.

Selection by the instructor of topics of interest will be very important, because normally I've found that I can't cover anywhere near all of this material in a semester. A reasonable choice for a first try might be to begin with Chapter 2 (recursive algorithms) which contains lots of motivation. Then, as new ideas are needed in Chapter 2, one might delve into the appropriate sections of Chapter 1 to get the concepts and techniques well in hand. After Chapter 2, Chapter 4, on number theory, discusses material that is extremely attractive, and surprisingly pure and applicable at the same time. Chapter 5 would be next, since the foundations would then all be in place. Finally, material from Chapter 3, which is rather independent of the rest of the book, but is strongly connected to combinatorial algorithms in general, might be studied as time permits.

Throughout the book there are opportunities to ask students to write programs and get them running. These are not mentioned explicitly, with a few exceptions, but will be obvious when encountered. Students should all have the experience of writing, debugging, and using a program that is nontrivially recursive, for example. The concept of recursion is subtle and powerful, and is helped a lot by hands-on practice. Any of the algorithms of Chapter 2 would be suitable for this purpose. The recursive graph algorithms are particularly recommended since they are usually quite foreign to students' previous experience and therefore have great learning value.

In addition to the exercises that appear in this book, then, student assignments might consist of writing occasional programs, as well as delivering reports in class on assigned readings. The latter might be found among the references cited in the bibliographies in each chapter.

I am indebted first of all to the students on whom I worked out these ideas, and second to a number of colleagues for their helpful advice and friendly criticism. Among the latter I will mention Richard Brualdi, Daniel Kleitman, Albert Nijenhuis, Robert Tarjan and Alan Tucker. For the no-doubt-numerous shortcomings that remain, I accept full responsibility.

This book was typeset in TeX. To the extent that it's a delight to look at, thank TeX. For the deficiencies in its appearance, thank my limitations as a typesetter. It was, however, a pleasure for me to have had the chance to typeset my own book. My thanks to the Computer Science department of the University of Pennsylvania, and particularly to Aravind Joshi, for generously allowing me the use of TeX facilities.

Herbert S. Wilf

Chapter 0: What This Book Is About

0.1 Background

An algorithm is a method for solving a class of problems on a computer. The complexity of an algorithm is the cost, measured in running time, or storage, or whatever units are relevant, of using the algorithm to solve one of those problems.

This book is about algorithms and complexity, and so it is about methods for solving problems on computers and the costs (usually the running time) of using those methods.

Computing takes time. Some problems take a very long time, others can be done quickly. Some problems *seem* to take a long time, and then someone discovers a faster way to do them (a 'faster algorithm'). The study of the amount of computational effort that is needed in order to perform certain kinds of computations is the study of computational *complexity*.

Naturally, we would expect that a computing problem for which millions of bits of input data are required would probably take longer than another problem that needs only a few items of input. So the time complexity of a calculation is measured by expressing the running time of the calculation *as a function of* some measure of the amount of data that is needed to describe the problem to the computer.

For instance, think about this statement: 'I just bought a matrix inversion program, and it can invert an $n \times n$ matrix in just $1.2n^3$ minutes.' We see here a typical description of the complexity of a certain algorithm. The running time of the program is being given as a function of the size of the input matrix.

A faster program for the same job might run in $0.8n^3$ minutes for an $n \times n$ matrix. If someone were to make a really important discovery (see section 2.4), then maybe we could actually lower the exponent, instead of merely shaving the multiplicative constant. Thus, a program that would invert an $n \times n$ matrix in only $7n^{2.8}$ minutes would represent a striking improvement of the state of the art.

For the purposes of this book, a computation that is guaranteed to take at most cn^3 time for input of size n will be thought of as an 'easy' computation. One that needs at most n^{10} time is also easy. If a certain

1

calculation on an $n \times n$ matrix were to require 2^n minutes, then that would be a 'hard' problem. Naturally some of the computations that we are calling 'easy' may take a very long time to run, but still, from our present point of view the important distinction to maintain will be the polynomial time guarantee or lack of it.

The general rule is that if the running time is at most a polynomial function of the amount of input data, then the calculation is an easy one, otherwise it's hard.

Many problems in computer science are known to be easy. To convince someone that a problem is easy, it is enough to describe a fast method for solving that problem. To convince someone that a problem is hard is hard, because you will have to prove to them that it is *impossible* to find a fast way of doing the calculation. It will *not* be enough to point to a particular algorithm and to lament its slowness. After all, *that* algorithm may be slow, but maybe there's a faster way.

Matrix inversion is easy. The familiar Gaussian elimination method can invert an $n \times n$ matrix in time at most cn^3.

To give an example of a hard computational problem we have to go far afield. One interesting one is called the 'tiling problem.' Suppose* we are given infinitely many identical floor tiles, each shaped like a regular hexagon. Then we can tile the whole plane with them, *i.e.*, we can cover the plane with no empty spaces left over. This can also be done if the tiles are identical rectangles, but not if they are regular pentagons.

In Fig. 0.1 we show a tiling of the plane by identical rectangles, and in Fig. 0.2 is a tiling by regular hexagons.

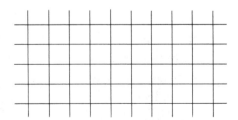

Fig. 0.1: Tiling with rectangles

---

* See, for instance, Martin Gardner's article in *Scientific American*, January 1977, pp. 110-121.

Fig. 0.2: Tiling with hexagons

That raises a number of theoretical and computational questions. One computational question is this. Suppose we are given a certain polygon, not necessarily regular and not necessarily convex, and suppose we have infinitely many identical tiles in that shape. Can we or can we not succeed in tiling the whole plane?

That elegant question has been *proved** to be computationally unsolvable. In other words, not only do we not know of any fast way to solve that problem on a computer, it has been *proved* that there isn't *any* way to do it, so even looking for an algorithm would be fruitless. That doesn't mean that the question is hard for every polygon. Hard problems can have easy instances. What has been proved is that no single method exists that can guarantee that it will decide this question for every polygon.

The fact that a computational *problem* is hard doesn't mean that every instance of it has to be hard. The *problem* is hard because we cannot devise an algorithm for which we can give a *guarantee* of fast performance for *all* instances.

Notice that the amount of input data to the computer in this example is quite small. All we need to input is the shape of the basic polygon. Yet not only is it impossible to devise a fast algorithm for this problem, it has been proved impossible to devise any algorithm at all that is guaranteed to terminate with a Yes/No answer after finitely many steps. That's *really* hard!

0.2 Hard vs. easy problems

Let's take a moment more to say in another way exactly what we mean by an 'easy' computation vs. a 'hard' one.

Think of an algorithm as being a little box that can solve a certain class

---

* R. Berger, The undecidability of the domino problem, *Memoirs Amer. Math. Soc.* **66** (1966), Amer. Math. Soc., Providence, RI of computational problems. Into the box goes a description of a particular problem in that class, and then, after a certain amount of time, or of computational effort, the answer appears.

A 'fast' algorithm is one that carries a guarantee of fast performance. Here are some examples.

Example 1. *It is guaranteed that if the input problem is described with B bits of data, then an answer will be output after at most $6B^3$ minutes.*

Example 2. *It is guaranteed that every problem that can be input with B bits of data will be solved in at most $0.7B^{15}$ seconds.*

A performance guarantee, like the two above, is sometimes called a 'worst-case complexity estimate,' and it's easy to see why. If we have an algorithm that will, for example, sort any given sequence of numbers into ascending order of size (see section 2.2) it may find that some sequences are easier to sort than others.

For instance, the sequence 1, 2, 7, 11, 10, 15, 20 is nearly in order already, so our algorithm might, if it takes advantage of the near-order, sort it very rapidly. Other sequences might be a lot harder for it to handle, and might therefore take more time.

So in some problems whose input bit string has B bits the algorithm might operate in time $6B$, and on others it might need, say, $10B \log B$ time units, and for still other problem instances of length B bits the algorithm might need $5B^2$ time units to get the job done.

Well then, what would the warranty card say? It would have to pick out the worst possibility, otherwise the guarantee wouldn't be valid. It would assure a user that if the input problem instance can be described by B bits, then an answer will appear after at most $5B^2$ time units. Hence a performance guarantee is equivalent to an estimation of the worst possible scenario: the longest possible calculation that might ensue if B bits are input to the program.

Worst-case bounds are the most common kind, but there are other kinds of bounds for running time. We might give an *average* case bound instead (see section 5.7). That wouldn't *guarantee* performance no worse than so-and-so; it would state that if the performance is averaged over all possible input bit strings of B bits, then the average amount of computing time will be so-and-so (as a function of B).

Now let's talk about the difference between easy and hard computational problems and between fast and slow algorithms.

A warranty that would *not* guarantee 'fast' performance would contain some function of B that grows faster than *any* polynomial. Like e^B, for instance, or like $2^{\sqrt{B}}$, etc. *It is the polynomial time vs. not necessarily polynomial time guarantee that makes the difference between the easy and the hard classes of problems, or between the fast and the slow algorithms.*

It is highly desirable to work with algorithms such that we can give a performance guarantee for their running time that is at most a polynomial function of the number of bits of input.

An algorithm is *slow* if, whatever polynomial P we think of, there exist arbitrarily large values of B, and input data strings of B bits, that cause the algorithm to do more than $P(B)$ units of work.

A computational problem is *tractable* if there is a fast algorithm that will do all instances of it.

A computational problem is *intractable* if it can be proved that there is no fast algorithm for it.

Example 3. Here is a familiar computational problem and a method, or algorithm, for solving it. Let's see if the method has a polynomial time guarantee or not.

The problem is this. Let n be a given integer. We want to find out if n is *prime*. The method that we choose is the following. For each integer $m = 2, 3, \ldots, \lfloor \sqrt{n} \rfloor$ we ask if m divides (evenly into) n. If all of the answers are 'No,' then we declare n to be a prime number, else it is composite.

We will now look at the *computational complexity* of this algorithm. That means that we are going to find out how much work is involved in doing the test. For a given integer n the work that we have to do can be measured in units of divisions of a whole number by another whole number. In those units, we obviously will do about $\sqrt{n}$ units of work.

It seems as though this is a tractable problem, because, after all, $\sqrt{n}$ is of polynomial growth in n. For instance, we do less than n units of work, and that's certainly a polynomial in n, isn't it? So, according to our definition of fast and slow algorithms, the distinction was made on the basis of polynomial vs. faster-than-polynomial growth of the work done with the problem size, and therefore this problem must be easy. Right? Well no,

5

not really.

Reference to the distinction between fast and slow methods will show that we have to measure the amount of work done *as a function of the number of bits of input to the problem.* In this example, n is not the number of bits of input. For instance, if $n = 59$, we don't need 59 bits to describe n, but only 6. In general, the number of binary digits in the bit string of an integer n is close to $\log_2 n$.

So in the problem of this example, testing the primality of a given integer n, the length of the input bit string B is about $\log_2 n$. Seen in this light, the calculation suddenly seems very long. A string consisting of a mere $\log_2 n$ 0's and 1's has caused our mighty computer to do about $\sqrt{n}$ units of work.

If we express the amount of work done as a function of B, we find that the complexity of this calculation is approximately $2^{B/2}$, and that grows much faster than any polynomial function of B.

Therefore, the method that we have just discussed for testing the primality of a given integer is slow. See chapter 4 for further discussion of this problem. At the present time no one has found a fast way to test for primality, nor has anyone proved that there isn't a fast way. Primality testing belongs to the (well-populated) class of seemingly, but not provably, intractable problems. ■

In this book we will deal with some easy problems and some seemingly hard ones. It's the 'seemingly' that makes things very interesting. These are problems for which no one has found a fast computer algorithm, but also, no one has proved the impossibility of doing so. It should be added that the entire area is vigorously being researched because of the attractiveness and the importance of the many unanswered questions that remain.

Thus, even though we just don't know many things that we'd like to know in this field , it isn't for lack of trying!

0.3 A preview

Chapter 1 contains some of the mathematical background that will be needed for our study of algorithms. It is not intended that reading this book or using it as a text in a course must necessarily begin with Chapter 1. It's probably a better idea to plunge into Chapter 2 directly, and then when particular skills or concepts are needed, to read the relevant portions of Chapter 1. Otherwise the definitions and ideas that are in that chapter may seem to be unmotivated, when in fact motivation in great quantity resides in the later chapters of the book.

Chapter 2 deals with recursive algorithms and the analyses of their complexities.

Chapter 3 is about a problem that seems as though it might be hard, but turns out to be easy, namely the network flow problem. Thanks to quite recent research, there are fast algorithms for network flow problems, and they have many important applications.

In Chapter 4 we study algorithms in one of the oldest branches of mathematics, the theory of numbers. Remarkably, the connections between this ancient subject and the most modern research in computer methods are very strong.

In Chapter 5 we will see that there is a large family of problems, including a number of very important computational questions, that are bound together by a good deal of structural unity. We don't know if they're hard or easy. We do know that we haven't found a fast way to do them yet, and most people suspect that they're hard. We also know that if any one of these problems is hard, then they all are, and if any one of them is easy, then they all are.

We hope that, having found out something about what people know and what people don't know, the reader will have enjoyed the trip through this subject and may be interested in helping to find out a little more.

Chapter 1: Mathematical Preliminaries

1.1 Orders of magnitude

In this section we're going to discuss the rates of growth of different functions and to introduce the five symbols of asymptotics that are used to describe those rates of growth. In the context of algorithms, the reason for this discussion is that we need a good language for the purpose of comparing the speeds with which different algorithms do the same job, or the amounts of memory that they use, or whatever other measure of the complexity of the algorithm we happen to be using.

Suppose we have a method of inverting square nonsingular matrices. How might we measure its speed? Most commonly we would say something like 'if the matrix is $n \times n$ then the method will run in time $16.8n^3$.' Then we would know that if a 100×100 matrix can be inverted, with this method, in 1 minute of computer time, then a 200×200 matrix would require $2^3 = 8$ times as long, or about 8 minutes. The constant '16.8' wasn't used at all in this example; only the fact that the labor grows as the third power of the matrix size was relevant.

Hence we need a language that will allow us to say that the computing time, as a function of n, grows 'on the order of n^3,' or 'at most as fast as n^3,' or 'at least as fast as $n^5 \log n$,' etc.

The new symbols that are used in the language of comparing the rates of growth of functions are the following five: 'o' (read 'is little oh of'), 'O' (read 'is big oh of'), 'Θ' (read 'is theta of'), '$\sim$' (read 'is asymptotically equal to' or, irreverently, as 'twiddles'), and 'Ω' (read 'is omega of').

Now let's explain what each of them means.

Let $f(x)$ and $g(x)$ be two functions of x. Each of the five symbols above is intended to compare the rapidity of growth of f and g. If we say that $f(x) = o(g(x))$, then informally we are saying that f grows more slowly than g does when x is very large. Formally, we state the

Definition. *We say that $f(x) = o(g(x))$ $(x \to \infty)$ if $\lim_{x \to \infty} f(x)/g(x)$ exists and is equal to 0.*

Here are some examples:

(a) $x^2 = o(x^5)$

8

(b) $\sin x = o(x)$

(c) $14.709\sqrt{x} = o(x/2 + 7\cos x)$

(d) $1/x = o(1)$ (?)

(e) $23\log x = o(x^{.02})$

We can see already from these few examples that sometimes it might be easy to prove that a '*o*' relationship is true and sometimes it might be rather difficult. Example (e), for instance, requires the use of L'Hospital's rule.

If we have two computer programs, and if one of them inverts $n \times n$ matrices in time $635n^3$ and if the other one does so in time $o(n^{2.8})$ then we know that *for all sufficiently large values of* n the performance guarantee of the second program will be superior to that of the first program. Of course, the first program might run faster on small matrices, say up to size $10,000 \times 10,000$. If a certain program runs in time $n^{2.03}$ and if someone were to produce another program for the same problem that runs in $o(n^2 \log n)$ time, then that second program would be an improvement, at least in the theoretical sense. The reason for the 'theoretical' qualification, once more, is that the second program would be known to be superior only if n were sufficiently large.

The second symbol of the asymptotics vocabulary is the '*O*.' When we say that $f(x) = O(g(x))$ we mean, informally, that f certainly doesn't grow at a faster rate than g. It might grow at the same rate or it might grow more slowly; both are possibilities that the '*O*' permits. Formally, we have the next

Definition. *We say that* $f(x) = O(g(x))$ $(x \to \infty)$ *if* $\exists C, x_0$ *such that* $|f(x)| < Cg(x)$ $(\forall x > x_0)$.

The qualifier '$x \to \infty$' will usually be omitted, since it will be understood that we will most often be interested in large values of the variables that are involved.

For example, it is certainly true that $\sin x = O(x)$, but even more can be said, namely that $\sin x = O(1)$. Also $x^3 + 5x^2 + 77\cos x = O(x^5)$ and $1/(1 + x^2) = O(1)$. Now we can see how the '*o*' gives more precise information than the '*O*,' for we can sharpen the last example by saying that $1/(1 + x^2) = o(1)$. This is sharper because not only does it tell us that the function is bounded when x is large, we learn that the function

9

actually approaches 0 as $x \to \infty$. This is typical of the relationship between O and o. It often happens that a 'O' result is sufficient for an application. However, that may not be the case, and we may need the more precise 'o' estimate.

The third symbol of the language of asymptotics is the 'Θ.'

Definition. *We say that $f(x) = \Theta(g(x))$ if there are constants $c_1 \neq 0$, $c_2 \neq 0$, x_0 such that for all $x > x_0$ it is true that $c_1 g(x) < f(x) < c_2 g(x)$.*

We might then say that f and g are of the same rate of growth, only the multiplicative constants are uncertain. Some examples of the 'Θ' at work are

$$(x+1)^2 = \Theta(3x^2)$$
$$(x^2 + 5x + 7)/(5x^3 + 7x + 2) = \Theta(1/x)$$
$$\sqrt{3 + \sqrt{2x}} = \Theta(x^{\frac{1}{4}})$$
$$(1 + 3/x)^x = \Theta(1).$$

The 'Θ' is much more precise than either the 'O' or the 'o.' If we know that $f(x) = \Theta(x^2)$, then we know that $f(x)/x^2$ stays between two nonzero constants for all sufficiently large values of x. The rate of growth of f is established: it grows quadratically with x.

The most precise of the symbols of asymptotics is the '$\sim$.' It tells us that not only do f and g grow at the same rate, but that in fact f/g approaches 1 as $x \to \infty$.

Definition. *We say that $f(x) \sim g(x)$ if $\lim_{x\to\infty} f(x)/g(x) = 1$.*

Here are some examples.

$$x^2 + x \sim x^2$$
$$(3x + 1)^4 \sim 81x^4$$
$$\sin 1/x \sim 1/x$$
$$(2x^3 + 5x + 7)/(x^2 + 4) \sim 2x$$
$$2^x + 7\log x + \cos x \sim 2^x$$

Observe the importance of getting the multiplicative constants exactly right when the '$\sim$' symbol is used. While it is true that $2x^2 = \Theta(x^2)$, it is not true that $2x^2 \sim x^2$. It is, by the way, also true that $2x^2 = \Theta(17x^2)$, but to make such an assertion is to use bad style since no more information is conveyed with the '17' than without it.

The last symbol in the asymptotic set that we will need is the 'Ω.' In a nutshell, 'Ω' is the negation of 'o.' That is to say, $f(x) = \Omega(g(x))$ means that it is *not* true that $f(x) = o(g(x))$. In the study of algorithms for computers, the 'Ω' is used when we want to express the thought that a certain calculation takes *at least* so-and-so long to do. For instance, we can multiply together two $n \times n$ matrices in time $O(n^3)$. Later on in this book we will see how to multiply two matrices even faster, in time $O(n^{2.81})$. People know of even faster ways to do that job, but one thing that we can be sure of is this: nobody will ever be able to write a matrix multiplication program that will multiply pairs $n \times n$ matrices with fewer than n^2 computational steps, because whatever program we write will have to look at the input data, and there are $2n^2$ entries in the input matrices.

Thus, a computing time of cn^2 is certainly a *lower bound* on the speed of any possible general matrix multiplication program. We might say, therefore, that the problem of multiplying two $n \times n$ matrices requires $\Omega(n^2)$ time.

The exact definition of the 'Ω' that was given above is actually rather delicate. We stated it as the negation of something. Can we rephrase it as a positive assertion? Yes, with a bit of work (see exercises 6 and 7 below). Since '$f = o(g)$' means that $f/g \to 0$, the symbol $f = \Omega(g)$ means that f/g does not approach zero. If we assume that g takes positive values only, which is usually the case in practice, then to say that f/g *does not* approach 0 is to say that $\exists \epsilon > 0$ and an infinite sequence of values of x, tending to ∞, along which $|f|/g > \epsilon$. So we don't have to show that $|f|/g > \epsilon$ *for all* large x, but only for *infinitely many* large x.

Definition. *We say that $f(x) = \Omega(g(x))$ if there is an $\epsilon > 0$ and a sequence $x_1, x_2, x_3, \ldots \to \infty$ such that $\forall j : |f(x_j)| > \epsilon g(x_j)$.*

Now let's introduce a hierarchy of functions according to their rates of growth when x is large. Among commonly occurring functions of x that grow without bound as $x \to \infty$, perhaps the slowest growing ones are functions like $\log \log x$ or maybe $(\log \log x)^{1.03}$ or things of that sort. It is certainly true that $\log \log x \to \infty$ as $x \to \infty$, but it takes its time about it. When $x = 1,000,000$, for example, $\log \log x$ has the value 2.6.

Just a bit faster growing than the 'snails' above is $\log x$ itself. After all, $\log(1,000,000) = 13.8$. So if we had a computer algorithm that could do n things in time $\log n$ and someone found another method that could do the same job in time $O(\log \log n)$, then the second method, other things being equal, would indeed be an improvement, but n might have to be extremely large before you would notice the improvement.

Next on the scale of rapidity of growth we might mention the powers of x. For instance, think about $x^{.01}$. It grows faster than $\log x$, although you wouldn't believe it if you tried to substitute a few values of x and to compare the answers (see exercise 1 at the end of this section).

How would we *prove* that $x^{.01}$ grows faster than $\log x$? By using L'Hospital's rule.

Example. Consider the limit of $x^{.01}/\log x$ for $x \to \infty$. As $x \to \infty$ the ratio assumes the indeterminate form ∞/∞, and it is therefore a candidate for L'Hospital's rule, which tells us that if we want to find the limit then we can differentiate the numerator, differentiate the denominator, and try again to let $x \to \infty$. If we do this, then instead of the original ratio, we find the ratio

$$.01x^{-.99}/(1/x) = .01x^{.01}$$

which obviously grows without bound as $x \to \infty$. Therefore the original ratio $x^{.01}/\log x$ also grows without bound. What we have proved, precisely, is that $\log x = o(x^{.01})$, and therefore in that sense we can say that $x^{.01}$ *grows faster than* $\log x$. ■

To continue up the scale of rates of growth, we meet $x^{.2}$, x, x^{15}, $x^{15} \log^2 x$, etc., and then we encounter functions that grow faster than *every* fixed power of x, just as $\log x$ grows slower than every fixed power of x.

Consider $e^{\log^2 x}$. Since this is the same as $x^{\log x}$ it will obviously grow faster than x^{1000}, in fact it will be larger than x^{1000} as soon as $\log x > 1000$, *i.e.*, as soon as $x > e^{1000}$ (don't hold your breath!).

Hence $e^{\log^2 x}$ is an example of a function that grows faster than every fixed power of x. Another such example is $e^{\sqrt{x}}$ (why?).

Definition. *A function that grows faster than x^a, for every constant a, but grows slower than c^x for every constant $c > 1$ is said to be of moderately exponential growth. More precisely, $f(x)$ is of moderately exponential* *growth if for every $a > 0$ we have $f(x) = \Omega(x^a)$ and for every $\epsilon > 0$ we have $f(x) = o((1 + \epsilon)^x)$.*

Beyond the range of moderately exponential growth are the functions that grow exponentially fast. Typical of such functions are $(1.03)^x$, 2^x, $x^9 7^x$, and so forth. Formally, we have the

Definition. *A function f is of exponential growth if there exists $c > 1$ such that $f(x) = \Omega(c^x)$ and there exists d such that $f(x) = O(d^x)$.*

If we clutter up a function of exponential growth with smaller functions then we will not change the fact that it is of exponential growth. Thus $e^{\sqrt{x}+2x}/(x^{49} + 37)$ remains of exponential growth, because e^{2x} is, all by itself, and it resists the efforts of the smaller functions to change its mind.

Beyond the exponentially growing functions there are functions that grow as fast as you might please. Like $n!$, for instance, which grows faster than c^n for every fixed constant c, and like 2^{n^2}, which grows much faster than $n!$. The growth ranges that are of the most concern to computer scientists are 'between' the very slowly, logarithmically growing functions and the functions that are of exponential growth. The reason is simple: if a computer algorithm requires more than an exponential amount of time to do its job, then it will probably not be used, or at any rate it will be used only in highly unusual circumstances. In this book, the algorithms that we will deal with all fall in this range.

Now we have discussed the various symbols of asymptotics that are used to compare the rates of growth of pairs of functions, and we have discussed the pecking order of rapidity of growth, so that we have a small catalogue of functions that grow slowly, medium-fast, fast, and super-fast. Next let's look at the growth of sums that involve elementary functions, with a view toward discovering the rates at which the sums grow.

Think about this one:

$$f(n) = \sum_{j=0}^{n} j^2 \tag{1.1.1}$$
$$= 1^2 + 2^2 + 3^2 + \cdots + n^2.$$

Thus, $f(n)$ is the sum of the squares of the first n positive integers. How fast does $f(n)$ grow when n is large?

Notice at once that among the n terms in the sum that defines $f(n)$, the biggest one is the last one, namely n^2. Since there are n terms in the sum and the biggest one is only n^2, it is certainly true that $f(n) = O(n^3)$, and even more, that $f(n) \le n^3$ for all $n \ge 1$.

Suppose we wanted more precise information about the growth of $f(n)$, such as a statement like $f(n) \sim ?$. How might we make such a better estimate?

The best way to begin is to visualize the sum in (1.1.1) as shown in Fig. 1.1.1.

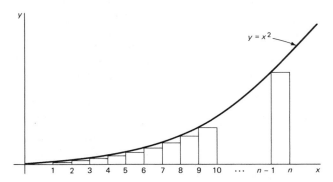

Fig. 1.1.1: How to overestimate a sum

In that figure we see the graph of the curve $y = x^2$, in the x-y plane. Further, there is a rectangle drawn over every interval of unit length in the range from $x = 1$ to $x = n$. The rectangles all lie *under* the curve. Consequently, the total area of all of the rectangles is smaller than the area under the curve, which is to say that

$$\sum_{j=1}^{n-1} j^2 \le \int_1^n x^2 \, dx$$

$$= (n^3 - 1)/3. \tag{1.1.2}$$

If we compare (1.1.2) and (1.1.1) we notice that we have proved that $f(n) \le ((n+1)^3 - 1)/3$.

Now we're going to get a *lower* bound on $f(n)$ in the same way. This time we use the setup in Fig. 1.1.2, where we again show the curve $y = x^2$, but this time we have drawn the rectangles so they lie *above* the curve.

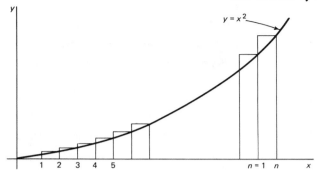

Fig. 1.1.2: How to underestimate a sum

From the picture we see immediately that

$$1^2 + 2^2 + \cdots + n^2 \geq \int_0^n x^2 \, dx$$
$$= n^3/3.$$

(1.1.3)

Now our function $f(n)$ has been bounded on both sides, rather tightly. What we know about it is that

$$\forall n \geq 1: \quad n^3/3 \leq f(n) \leq ((n+1)^3 - 1)/3.$$

From this we have immediately that $f(n) \sim n^3/3$, which gives us quite a good idea of the rate of growth of $f(n)$ when n is large. The reader will also have noticed that the '$\sim$' gives a much more satisfying estimate of growth than the 'O' does.

Let's formulate a general principle, for estimating the size of a sum, that will make estimates like the above for us without requiring us each time to visualize pictures like Figs. 1.1.1 and 1.1.2. The general idea is that when one is faced with estimating the rates of growth of sums, then one should try to compare the sums with integrals because they're usually easier to deal with.

Let a function $g(n)$ be defined for nonnegative integer values of n, and suppose that $g(n)$ is nondecreasing. We want to estimate the growth of the sum

$$G(n) = \sum_{j=1}^{n} g(j) \quad (n = 1, 2, \ldots).$$

(1.1.4)

Consider a diagram that looks exactly like Fig. 1.1.1 except that the curve that is shown there is now the curve $y = g(x)$. The sum of the areas of the

15

rectangles is exactly $G(n-1)$, while the area under the curve between 1 and n is $\int_1^n g(t)\,dt$. Since the rectangles lie wholly under the curve, their combined areas cannot exceed the area under the curve, and we have the inequality

$$G(n-1) \leq \int_1^n g(t)\,dt \qquad (n \geq 1). \tag{1.1.5}$$

On the other hand, if we consider Fig. 1.1.2, where the graph is once more the graph of $y = g(x)$, the fact that the combined areas of the rectangles is now *not less than* the area under the curve yields the inequality

$$G(n) \geq \int_0^n g(t)\,dt \qquad (n \geq 1). \tag{1.1.6}$$

If we combine (1.1.5) and (1.1.6) we find that we have completed the proof of

Theorem 1.1.1. *Let $g(x)$ be nondecreasing for nonnegative x. Then*

$$\int_0^n g(t)\,dt \leq \sum_{j=1}^n g(j) \leq \int_1^{n+1} g(t)\,dt. \tag{1.1.7}$$

The above theorem is capable of producing quite satisfactory estimates with rather little labor, as the following example shows.

Let $g(n) = \log n$ and substitute in (1.1.7). After doing the integrals, we obtain

$$n \log n - n \leq \sum_{j=1}^n \log j \leq (n+1) \log(n+1) - n. \tag{1.1.8}$$

We recognize the middle member above as $\log n!$, and therefore by exponentiation of (1.1.8) we have

$$\left(\frac{n}{e}\right)^n \leq n! \leq \frac{(n+1)^{n+1}}{e^n}. \tag{1.1.9}$$

This is rather a good estimate of the growth of $n!$, since the right member is only about ne times as large as the left member (why?), when n is large.

By the use of slightly more precise machinery one can prove a better estimate of the size of $n!$ that is called *Stirling's formula*, which is the statement that

$$x! \sim \left(\frac{x}{e}\right)^x \sqrt{2x\pi}. \tag{1.1.10}$$

Exercises for section 1.1

1. Calculate the values of $x^{.01}$ and of $\log x$ for $x = 10, 1000, 1{,}000{,}000$. Find a single value of $x > 10$ for which $x^{.01} > \log x$, and prove that your answer is correct.

2. Some of the following statements are true and some are false. Which are which?

    (a) $(x^2 + 3x + 1)^3 \sim x^6$

    (b) $(\sqrt{x} + 1)^3/(x^2 + 1) = o(1)$

    (c) $e^{1/x} = \Theta(1)$

    (d) $1/x \sim 0$

    (e) $x^3 (\log \log x)^2 = o(x^3 \log x)$

    (f) $\sqrt{\log x + 1} = \Omega(\log \log x)$

    (g) $\sin x = \Omega(1)$

    (h) $\cos x/x = O(1)$

    (i) $\int_4^x dt/t \sim \log x$

    (j) $\int_0^x e^{-t^2} dt = O(1)$

    (k) $\sum_{j \le x} 1/j^2 = o(1)$

    (l) $\sum_{j \le x} 1 \sim x$

3. Each of the three sums below defines a function of x. Beneath each sum there appears a list of five assertions about the rate of growth, as $x \to \infty$, of the function that the sum defines. In each case state which of the five choices, if any, are true (note: more than one choice may be true).

$$h_1(x) = \sum_{j \le x} \{1/j + 3/j^2 + 4/j^3\}$$

(i) $\sim \log x$ (ii) $= O(x)$ (iii) $\sim 2 \log x$ (iv) $= \Theta(\log x)$ (v) $= \Omega(1)$

$$h_2(x) = \sum_{j \le \sqrt{x}} \{\log j + j\}$$

(i) $\sim x/2$ (ii) $= O(\sqrt{x})$ (iii) $= \Theta(\sqrt{x} \log x)$ (iv) $= \Omega(\sqrt{x})$ (v) $= o(\sqrt{x})$

$$h_3(x) = \sum_{j \le \sqrt{x}} 1/\sqrt{j}$$

(i) $= O(\sqrt{x})$ (ii) $= \Omega(x^{1/4})$ (iii) $= o(x^{1/4})$ (iv) $\sim 2x^{1/4}$ (v) $= \Theta(x^{1/4})$

17

4. Of the five symbols of asymptotics $O, o, \sim, \Theta, \Omega$, which ones are *transitive* (*e.g.*, if $f = O(g)$ and $g = O(h)$, is $f = O(h)$?)?

5. The point of this exercise is that if f grows more slowly than g, then we can always find a third function h whose rate of growth is between that of f and of g. Precisely, prove the following: if $f = o(g)$ then there is a function h such that $f = o(h)$ and $h = o(g)$. Give an explicit construction for the function h in terms of f and g.

6. {This exercise is a warmup for exercise 7.} Below there appear several mathematical propositions. In each case, write a proposition that is the negation of the given one. Furthermore, in the negation, do not use the word 'not' or any negation symbols. In each case the question is, 'If this *isn't* true, then what *is* true?'

    (a) $\exists x > 0 \ni f(x) \neq 0$

    (b) $\forall x > 0, \ f(x) > 0$

    (c) $\forall x > 0, \ \exists \epsilon > 0 \ni f(x) < \epsilon$

    (d) $\exists x \neq 0 \ni \forall y < 0, \ f(y) < f(x)$

    (e) $\forall x \exists y \ni \forall z : g(x) < f(y)f(z)$

    (f) $\forall \epsilon > 0 \exists x \ni \forall y > x : f(y) < \epsilon$

Can you formulate a general method for negating such propositions? Given a proposition that contains '$\forall$,' '$\exists$,' '$\ni$,' what rule would you apply in order to negate the proposition and leave the result in positive form (containing no negation symbols or 'not's).

7. In this exercise we will work out the definition of the 'Ω.'

    (a) Write out the precise definition of the statement '$\lim_{x \to \infty} h(x) = 0$' (use '$\epsilon$'s).

    (b) Write out the negation of your answer to part (a), as a positive assertion.

    (c) Use your answer to part (b) to give a positive definition of the assertion '$f(x) \neq o(g(x))$,' and thereby justify the definition of the 'Ω' symbol that was given in the text.

8. Arrange the following functions in increasing order of their rates of growth, for large n. That is, list them so that each one is 'little oh' of its successor:

$$2^{\sqrt{n}}, e^{\log n^3}, n^{3.01}, 2^{n^2},$$

$$n^{1.6}, \log n^3 + 1, \sqrt{n!}, n^{3 \log n},$$

$$n^3 \log n, (\log \log n)^3, n^{.5} 2^n, (n+4)^{12}$$

9. Find a function $f(x)$ such that $f(x) = O(x^{1+\epsilon})$ is true for every $\epsilon > 0$, but for which it is not true that $f(x) = O(x)$.

10. Prove that the statement '$f(n) = O((2 + \epsilon)^n)$ for every $\epsilon > 0$' is equivalent to the statement '$f(n) = o((2 + \epsilon)^n)$ for every $\epsilon > 0$.'

1.2 Positional number systems

This section will provide a brief review of the representation of numbers in different bases. The usual decimal system represents numbers by using the digits $0, 1, \ldots, 9$. For the purpose of representing whole numbers we can imagine that the powers of 10 are displayed before us like this:

$$\ldots, 100000, 10000, 1000, 100, 10, 1.$$

Then, to represent an integer we can specify how many copies of each power of 10 we would like to have. If we write 237, for example, then that means that we want 2 100's and 3 10's and 7 1's.

In general, if we write out the string of digits that represents a number in the decimal system, as $d_m d_{m-1} \cdots d_1 d_0$, then the number that is being represented by that string of digits is

$$n = \sum_{i=0}^{m} d_i 10^i.$$

Now let's try the *binary system*. Instead of using 10's we're going to use 2's. So we imagine that the powers of 2 are displayed before us, as

$$\ldots, 512, 256, 128, 64, 32, 16, 8, 4, 2, 1.$$

To represent a number we will now specify how many copies of each power of 2 we would like to have. For instance, if we write 1101, then we want an 8, a 4 and a 1, so this must be the decimal number 13. We will write

$$(13)_{10} = (1101)_2$$

to mean that the number 13, in the base 10, is the same as the number 1101, in the base 2.

In the binary system (base 2) the only digits we will ever need are 0 and 1. What that means is that if we use only 0's and 1's then we can represent every number n in exactly one way. The unique representation of every number, is, after all, what we must expect and demand of any proposed system.

Let's elaborate on this last point. If we were allowed to use more digits than just 0's and 1's then we would be able to represent the number $(13)_{10}$ as a binary number in a whole lot of ways. For instance, we might make the mistake of allowing digits 0, 1, 2, 3. Then 13 would be representable by $3 \cdot 2^2 + 1 \cdot 2^0$ or by $2 \cdot 2^2 + 2 \cdot 2^1 + 1 \cdot 2^0$ etc.

So if we were to allow too *many* different digits, then numbers would be representable in more than one way by a string of digits.

If we were to allow *too few* different digits then we would find that some numbers have no representation at all. For instance, if we were to use the decimal system with only the digits $0, 1, \ldots, 8$, then infinitely many numbers would not be able to be represented, so we had better keep the 9's.

The general proposition is this.

Theorem 1.2.1. *Let $b > 1$ be a positive integer (the 'base'). Then every positive integer n can be written in one and only one way in the form*

$$n = d_0 + d_1 b + d_2 b^2 + d_3 b^3 + \cdots$$

if the digits $d_0, d_1, \ldots$ lie in the range $0 \le d_i \le b - 1$, for all i.

Remark: The theorem says, for instance, that in the base 10 we need the digits 0, 1, 2, ..., 9, in the base 2 we need only 0 and 1, in the base 16 we need sixteen digits, etc.

Proof of the theorem: If b is fixed, the proof is by induction on n, the number being represented. Clearly the number 1 can be represented in one and only one way with the available digits (why?). Suppose, inductively, that every integer $1, 2, \ldots, n - 1$ is uniquely representable. Now consider the integer n. Define $d = n \bmod b$. Then d is one of the b permissible digits. By induction, the number $n' = (n - d)/b$ is uniquely representable, say

$$\frac{n - d}{b} = d_0 + d_1 b + d_2 b^2 + \ldots$$

Then clearly,

$$n = d + \frac{n-d}{b}b$$
$$= d + d_0 b + d_1 b^2 + d_2 b^3 + \ldots$$

is a representation of n that uses only the allowed digits.

Finally, suppose that n has some other representation in this form also. Then we would have

$$n = a_0 + a_1 b + a_2 b^2 + \ldots$$
$$= c_0 + c_1 b + c_2 b^2 + \ldots$$

Since a_0 and c_0 are both equal to $n \bmod b$, they are equal to each other. Hence the number $n' = (n - a_0)/b$ has two different representations, which contradicts the inductive assumption, since we have assumed the truth of the result for all $n' < n$. ■

The bases b that are the most widely used are, aside from 10, 2 ('binary system'), 8 ('octal system') and 16 ('hexadecimal system').

The binary system is extremely simple because it uses only two digits. This is very convenient if you're a computer or a computer designer, because the digits can be determined by some component being either 'on' (digit 1) or 'off' (digit 0). The binary digits of a number are called its *bits* or its *bit string*.

The octal system is popular because it provides a good way to remember and deal with the long bit strings that the binary system creates. According to the theorem, in the octal system the digits that we need are $0, 1, \ldots, 7$. For instance,

$$(735)_8 = (477)_{10}.$$

The captivating feature of the octal system is the ease with which we can convert between octal and binary. If we are given the bit string of an integer n, then to convert it to octal, all we have to do is to group the bits together in groups of three, starting with the least significant bit, then convert each group of three bits, independently of the others, into a single octal digit. Conversely, if the octal form of n is given, then the binary form is obtainable by converting each octal digit independently into the three bits that represent it in the binary system.

For example, given $(1101100101)_2$. To convert this binary number to octal, we group the bits in threes,

$$(1)(101)(100)(101)$$

starting from the right, and then we convert each triple into a single octal digit, thereby getting

$$(1101100101)_2 = (1545)_8.$$

If you're a working programmer it's very handy to use the shorter octal strings to remember, or to write down, the longer binary strings, because of the space saving, coupled with the ease of conversion back and forth.

The hexadecimal system (base 16) is like octal, only more so. The conversion back and forth to binary now uses groups of *four* bits, rather than three. In hexadecimal we will need, according to the theorem above, 16 digits. We have handy names for the first 10 of these, but what shall we call the 'digits 10 through 15'? The names that are conventionally used for them are 'A,' 'B,'...,'F.' We have, for example,

$$
\begin{aligned}
(A52C)_{16} &= 10(4096) + 5(256) + 2(16) + 12 \\
&= (42284)_{10} \\
&= (1010)_2(0101)_2(0010)_2(1100)_2 \\
&= (1010010100101100)_2 \\
&= (1)(010)(010)(100)(101)(100) \\
&= (122454)_8.
\end{aligned}
$$

Exercises for section 1.2

1. Prove that conversion from octal to binary is correctly done by converting each octal digit to a binary triple and concatenating the resulting triples. Generalize this theorem to other pairs of bases.

2. Carry out the conversions indicated below.
    (a) $(737)_{10} = (?)_3$
    (b) $(101100)_2 = (?)_{16}$
    (c) $(3377)_8 = (?)_{16}$

(d) $(ABCD)_{16} = (?)_{10}$

(e) $(BEEF)_{16} = (?)_8$

3. Write a procedure *convert* (*n, b*:integer, *digitstr*:string), that will find the string of digits that represents n in the base b.

1.3 Manipulations with series

In this section we will look at operations with power series, including multiplying them and finding their sums in simple form. We begin with a little catalogue of some power series that are good to know. First we have the finite geometric series

$$(1 - x^n)/(1 - x) = 1 + x + x^2 + \cdots + x^{n-1}. \qquad (1.3.1)$$

This equation is valid certainly for all $x \neq 1$, and it remains true when $x = 1$ also if we take the limit indicated on the left side.

Why is (1.3.1) true? Just multiply both sides by $1 - x$ to clear of fractions. The result is

$$1 - x^n = (1 + x + x^2 + x^3 + \cdots + x^{n-1})(1 - x)$$
$$= (1 + x + x^2 + \cdots + x^{n-1}) - (x + x^2 + x^3 + \cdots + x^n)$$
$$= 1 - x^n$$

and the proof is finished.

Now try this one. What is the value of the sum

$$\sum_{j=0}^{9} 3^j ?$$

Observe that we are looking at the right side of (1.3.1) with $x = 3$. Therefore the answer is $(3^{10} - 1)/2$. Try to get used to the idea that *a series in powers of x becomes a number if x is replaced by a number*, and if we know a formula for the sum of the series then we know the number that it becomes.

Here are some more series to keep in your zoo. A parenthetical remark like '$(|x| < 1)$' shows the set of values of x for which the series converges.

$$\sum_{k=0}^{\infty} x^k = 1/(1 - x) \qquad (|x| < 1) \qquad (1.3.2)$$

$$e^x = \sum_{m=0}^{\infty} x^m/m! \tag{1.3.3}$$

$$\sin x = \sum_{r=0}^{\infty} (-1)^r x^{2r+1}/(2r+1)! \tag{1.3.4}$$

$$\cos x = \sum_{s=0}^{\infty} (-1)^s x^{2s}/(2s)! \tag{1.3.5}$$

$$\log\left(1/(1-x)\right) = \sum_{j=1}^{\infty} x^j/j \qquad (|x| < 1) \tag{1.3.6}$$

Can you find a simple form for the sum (the logarithms are 'natural')

$$1 + \log 2 + (\log 2)^2/2! + (\log 2)^3/3! + \cdots?$$

Hint: Look at (1.3.3), and replace x by $\log 2$.

Aside from merely substituting values of x into known series, there are many other ways of using known series to express sums in simple form. Let's think about the sum

$$1 + 2 \cdot 2 + 3 \cdot 4 + 4 \cdot 8 + 5 \cdot 16 + \cdots + N2^{N-1}. \tag{1.3.7}$$

We are reminded of the finite geometric series (1.3.1), but (1.3.7) is a little different because of the multipliers $1, 2, 3, 4, \ldots, N$.

The trick is this. When confronted with a series that is similar to, but not identical with, a known series, write down the known series as an equation, with the series on one side and its sum on the other. Even though the unknown series involves a particular value of x, in this case $x = 2$, keep the known series with its variable unrestricted. Then reach for an appropriate tool that will be applied to both sides of that equation, and whose result will be that the known series will have been changed into the one whose sum we needed.

In this case, since (1.3.7) reminds us of (1.3.1), we'll begin by writing down (1.3.1) again,

$$(1 - x^n)/(1 - x) = 1 + x + x^2 + \cdots + x^{n-1} \tag{1.3.8}$$

Don't replace x by 2 yet, just walk up to the equation (1.3.8) carrying your tool kit and ask what kind of surgery you could do to *both sides of* (1.3.8) that would be helpful in evaluating the unknown (1.3.7).

We are going to reach into our tool kit and pull out '$\frac{d}{dx}$.' In other words, we are going to *differentiate* (1.3.8). The reason for choosing differentiation is that it will put the missing multipliers $1, 2, 3, \ldots, N$ into (1.3.8). After differentiation, (1.3.8) becomes

$$1 + 2x + 3x^2 + 4x^3 + \cdots + (n-1)x^{n-2} = \frac{1 - nx^{n-1} + (n-1)x^n}{(1-x)^2}. \quad (1.3.9)$$

Now it's easy. To evaluate the sum (1.3.7), all we have to do is to substitute $x = 2$, $n = N + 1$ in (1.3.9), to obtain, after simplifying the right-hand side,

$$1 + 2 \cdot 2 + 3 \cdot 4 + 4 \cdot 8 + \cdots + N2^{N-1} = 1 + (N-1)2^N. \quad (1.3.10)$$

Next try this one:

$$\frac{1}{2 \cdot 3^2} + \frac{1}{3 \cdot 3^3} + \cdots \quad (1.3.11)$$

If we rewrite the series using summation signs, it becomes

$$\sum_{j=2}^{\infty} \frac{1}{j \cdot 3^j}.$$

Comparison with the series zoo shows great resemblance to the species (1.3.6). In fact, if we put $x = 1/3$ in (1.3.6) it tells us that

$$\sum_{j=1}^{\infty} \frac{1}{j \cdot 3^j} = \log(3/2). \quad (1.3.12)$$

The desired sum (1.3.11) is the result of dropping the term with $j = 1$ from (1.3.12), which shows that the sum in (1.3.11) is equal to $\log(3/2) - 1/3$.

In general, suppose that $f(x) = \sum a_n x^n$ is some series that we know. Then $\sum na_n x^{n-1} = f'(x)$ and $\sum na_n x^n = xf'(x)$. In other words, if the nth coefficient is multiplied by n, then the function changes from f to $(x\frac{d}{dx})f$. If we apply the rule again, we find that multiplying the nth coefficient of a power series by n^2 changes the sum from f to $(x\frac{d}{dx})^2 f$. That is,

$$\sum_{j=0}^{\infty} j^2 x^j / j! = (x\frac{d}{dx})(x\frac{d}{dx})e^x$$

$$= (x\frac{d}{dx})(xe^x)$$

$$= (x^2 + x)e^x.$$

Similarly, multiplying the nth coefficient of a power series by n^p will change the sum from $f(x)$ to $(x\frac{d}{dx})^p f(x)$, but that's not all. What happens if we multiply the coefficient of x^n by, say, $3n^2 + 2n + 5$? If the sum previously was $f(x)$, then it will be changed to $\{3(x\frac{d}{dx})^2 + 2(x\frac{d}{dx}) + 5\}f(x)$. The sum

$$\sum_{j=0}^{\infty}(2j^2 + 5)x^j$$

is therefore equal to $\{2(x\frac{d}{dx})^2 + 5\}\{1/(1-x)\}$, and after doing the differentiations we find the answer in the form $(7x^2 - 8x + 5)/(1-x)^3$.

Here is the general rule: if $P(x)$ is any polynomial then

$$\sum_j P(j)a_j x^j = P(x\frac{d}{dx})\{\sum_j a_j x^j\}. \tag{1.3.13}$$

Exercises for section 1.3

1. Find simple, explicit formulas for the sums of each of the following series.
   (a) $\sum_{j\geq 3}\log 6^j/j!$
   (b) $\sum_{m>1}(2m + 7)/5^m$
   (c) $\sum_{j=0}^{19}(j/2^j)$
   (d) $1 - x/2! + x^2/4! - x^3/6! + \cdots$
   (e) $1 - 1/3^2 + 1/3^4 - 1/3^6 + \cdots$
   (f) $\sum_{m=2}^{\infty}(m^2 + 3m + 2)/m!$
2. Explain why $\sum_{r\geq 0}(-1)^r\pi^{2r+1}/(2r + 1)! = 0$.
3. Find the coefficient of t^n in the series expansion of each of the following functions about $t = 0$.
   (a) $(1 + t + t^2)e^t$
   (b) $(3t - t^2)\sin t$
   (c) $(t + 1)^2/(t - 1)^2$

1.4 Recurrence relations

A recurrence relation is a formula that permits us to compute the members of a sequence one after another, starting with one or more given values.

Here is a small example. Suppose we are to find an infinite sequence of numbers $x_0, x_1, \ldots$ by means of

$$x_{n+1} = cx_n \qquad (n \geq 0; \ x_0 = 1). \tag{1.4.1}$$

This relation tells us that $x_1 = cx_0$, and $x_2 = cx_1$, etc., and furthermore that $x_0 = 1$. It is then clear that $x_1 = c$, $x_2 = c^2, \ldots, x_n = c^n, \ldots$

We say that the *solution* of the recurrence relation (= 'difference equation') (1.4.1) is given by $x_n = c^n$ for all $n \geq 0$. Equation (1.4.1) is a *first-order* recurrence relation because a new value of the sequence is computed from just one preceding value (*i.e.*, x_{n+1} is obtained solely from x_n, and does not involve x_{n-1} or any earlier values).

Observe the format of the equation (1.4.1). The parenthetical remarks are essential. The first one '$n \geq 0$' tells us for what values of n the recurrence formula is valid, and the second one '$x_0 = 1$' gives the starting value. If one of these is missing, the solution may not be uniquely determined. The recurrence relation

$$x_{n+1} = x_n + x_{n-1} \tag{1.4.2}$$

needs two starting values in order to 'get going,' but it is missing both of those starting values and the range of n. Consequently (1.4.2) (which is a second-order recurrence) does not uniquely determine the sequence.

The situation is rather similar to what happens in the theory of ordinary differential equations. There, if we omit initial or boundary values, then the solutions are determined only up to arbitrary constants.

Beyond the simple (1.4.1), the next level of difficulty occurs when we consider a first-order recurrence relation with a variable multiplier, such as

$$x_{n+1} = b_{n+1}x_n \qquad (n \geq 0; \ x_0 \text{ given}). \tag{1.4.3}$$

Now $\{b_1, b_2, \ldots\}$ is a given sequence, and we are being asked to find the unknown sequence $\{x_1, x_2, \ldots\}$.

In an easy case like this we can write out the first few x's and then guess the answer. We find, successively, that $x_1 = b_1x_0$, then $x_2 = b_2x_1 = b_2b_1x_0$ and $x_3 = b_3x_2 = b_3b_2b_1x_0$ etc. At this point we can guess that the solution is

$$x_n = \{\prod_{i=1}^{n} b_i\}x_0 \qquad (n = 0, 1, 2, \ldots). \tag{1.4.4}$$

Since that wasn't hard enough, we'll raise the ante a step further. Suppose we want to solve the first-order *inhomogeneous* (because $x_n = 0$ for all n is not a solution) recurrence relation

$$x_{n+1} = b_{n+1}x_n + c_{n+1} \qquad (n \geq 0;\ x_0\ \text{given}). \qquad (1.4.5)$$

Now we are being given two sequences $b_1, b_2, \ldots$ and $c_1, c_2, \ldots$, and we want to find the x's. Suppose we follow the strategy that has so far won the game, that is, writing down the first few x's and trying to guess the pattern. Then we would find that $x_1 = b_1 x_0 + c_1$, $x_2 = b_2 b_1 x_0 + b_2 c_1 + c_2$, and we would probably tire rapidly.

Here is a somewhat more orderly approach to (1.4.5). Though no approach will avoid the unpleasant form of the general answer, the one that we are about to describe at least gives a method that is much simpler than the guessing strategy, for many examples that arise in practice. In this book we are going to run into several equations of the type of (1.4.5), so a unified method will be a definite asset.

The first step is to define a new unknown function as follows. Let

$$x_n = b_1 b_2 \cdots b_n y_n \qquad (n \geq 1;\ x_0 = y_0) \qquad (1.4.6)$$

define a new unknown sequence $y_1, y_2, \ldots$ Now substitute for x_n in (1.4.5), getting

$$b_1 b_2 \cdots b_{n+1} y_{n+1} = b_{n+1} b_1 b_2 \cdots b_n y_n + c_{n+1}.$$

We notice that the coefficients of y_{n+1} and of y_n are the same, and so we divide both sides by that coefficient. The result is the equation

$$y_{n+1} = y_n + d_{n+1} \qquad (n \geq 0;\ y_0\ \text{given}) \qquad (1.4.7)$$

where we have written $d_{n+1} = c_{n+1}/(b_1 \cdots b_{n+1})$. Notice that the d's are *known*.

We haven't yet solved the recurrence relation. We have only changed to a new unknown function that satisfies a simpler recurrence (1.4.7). Now the solution of (1.4.7) is quite simple, because it says that each y is obtained from its predecessor by adding the next one of the d's. It follows that

$$y_n = y_0 + \sum_{j=1}^{n} d_j \qquad (n \geq 0).$$

We can now use (1.4.6) to reverse the change of variables to get back to the original unknowns $x_0, x_1, \ldots$, and find that

$$x_n = (b_1 b_2 \cdots b_n)\{x_0 + \sum_{j=1}^{n} d_j\} \qquad (n \geq 1). \qquad (1.4.8)$$

It is not recommended that the reader memorize the solution that we have just obtained. It *is* recommended that the method by which the solution was found be mastered. It involves

(a) make a change of variables that leads to a new recurrence of the form (1.4.6), then

(b) solve that one by summation and

(c) go back to the original unknowns.

As an example, consider the first-order equation

$$x_{n+1} = 3x_n + n \qquad (n \geq 0; \; x_0 = 0). \qquad (1.4.9)$$

The winning change of variable, from (1.4.6), is to let $x_n = 3^n y_n$. After substituting in (1.4.9) and simplifying, we find

$$y_{n+1} = y_n + n/3^{n+1} \qquad (n \geq 0; \; y_0 = 0).$$

Now by summation,

$$y_n = \sum_{j=1}^{n-1} j/3^{j+1} \qquad (n \geq 0).$$

Finally, since $x_n = 3^n y_n$ we obtain the solution of (1.4.9) in the form

$$x_n = 3^n \sum_{j=1}^{n-1} j/3^{j+1} \qquad (n \geq 0). \qquad (1.4.10)$$

This is quite an explicit answer, but the summation can, in fact, be completely removed by the same method that you used to solve exercise 1(c) of section 1.3 (try it!).

That pretty well takes care of first-order recurrence relations of the form $x_{n+1} = b_{n+1} x_n + c_{n+1}$, and it's time to move on to linear second order (homogeneous) recurrence relations with constant coefficients. These are of the form

$$x_{n+1} = ax_n + bx_{n-1} \qquad (n \geq 1; \; x_0 \text{ and } x_1 \text{ given}). \qquad (1.4.11)$$

If we think back to *differential* equations of second-order with constant coefficients, we recall that there are always solutions of the form $y(t) = e^{\alpha t}$ where α is constant. Hence the road to the solution of such a differential equation begins by trying a solution of that form and seeing what the constant or constants α turn out to be.

Analogously, equation (1.4.11) calls for a trial solution of the form $x_n = \alpha^n$. If we substitute $x_n = \alpha^n$ in (1.4.11) and cancel a common factor of α^{n-1} we obtain a quadratic equation for α, namely

$$\alpha^2 = a\alpha + b. \tag{1.4.12}$$

'Usually' this quadratic equation will have two distinct roots, say α_+ and α_-, and then the general solution of (1.4.11) will look like

$$x_n = c_1 \alpha_+^n + c_2 \alpha_-^n \qquad (n = 0, 1, 2, \ldots). \tag{1.4.13}$$

The constants c_1 and c_2 will be determined so that x_0, x_1 have their assigned values.

Example. The recurrence for the Fibonacci numbers is

$$F_{n+1} = F_n + F_{n-1} \qquad (n \geq 1; \; F_0 = F_1 = 1). \tag{1.4.14}$$

Following the recipe that was described above, we look for a solution in the form $F_n = \alpha^n$. After substituting in (1.4.14) and cancelling common factors we find that the quadratic equation for α is, in this case, $\alpha^2 = \alpha + 1$.

If we denote the two roots by $\alpha_+ = (1 + \sqrt{5})/2$ and $\alpha_- = (1 - \sqrt{5})/2$, then the general solution to the Fibonacci recurrence has been obtained, and it has the form (1.4.13). It remains to determine the constants c_1, c_2 from the initial conditions $F_0 = F_1 = 1$.

From the form of the general solution we have $F_0 = 1 = c_1 + c_2$ and $F_1 = 1 = c_1 \alpha_+ + c_2 \alpha_-$. If we solve these two equations in the two unknowns c_1, c_2 we find that $c_1 = \alpha_+/\sqrt{5}$ and $c_2 = -\alpha_-/\sqrt{5}$. Finally, we substitute these values of the constants into the form of the general solution, and obtain an explicit formula for the nth Fibonacci number,

$$F_n = \frac{1}{\sqrt{5}} \left\{ \left(\frac{1 + \sqrt{5}}{2} \right)^{n+1} - \left(\frac{1 - \sqrt{5}}{2} \right)^{n+1} \right\} \qquad (n = 0, 1, \ldots). \tag{1.4.15}$$

The Fibonacci numbers are in fact $1, 1, 2, 3, 5, 8, 13, 21, 34, \ldots$ It isn't even obvious that the formula (1.4.15) gives integer values for the F_n's. The reader should check that the formula indeed gives the first few F_n's correctly.

Just to exercise our newly acquired skills in asymptotics, let's observe that since $(1 + \sqrt{5})/2 > 1$ and $|(1 - \sqrt{5})/2| < 1$, it follows that when n is large we have

$$F_n \sim ((1 + \sqrt{5})/2)^{n+1}/\sqrt{5}.$$

■

The process of looking for a solution in a certain form, namely in the form α^n, is subject to the same kind of special treatment, in the case of repeated roots, that we find in differential equations. Corresponding to a *double* root α of the associated quadratic equation $\alpha^2 = a\alpha + b$ we would find two independent solutions α^n and $n\alpha^n$, so the general solution would be in the form $\alpha^n(c_1 + c_2 n)$.

Example. Consider the recurrence

$$x_{n+1} = 2x_n - x_{n-1} \qquad (n \geq 1; \ x_0 = 1; \ x_1 = 5). \tag{1.4.16}$$

If we try a solution of the type $x_n = \alpha^n$, then we find that α satisfies the quadratic equation $\alpha^2 = 2\alpha - 1$. Hence the 'two' roots are 1 and 1. The general solution is $x_n = 1^n(c_1 + nc_2) = c_1 + c_2 n$. After inserting the given initial conditions, we find that

$$x_0 = 1 = c_1; \ x_1 = 5 = c_1 + c_2$$

If we solve for c_1 and c_2 we obtain $c_1 = 1$, $c_2 = 4$, and therefore the complete solution of the recurrence (1.4.16) is given by $x_n = 4n + 1$. ■

Now let's look at recurrent *inequalities*, like this one:

$$x_{n+1} \leq x_n + x_{n-1} + n^2 \quad (n \geq 1; \ x_0 = 0; \ x_1 = 0). \tag{1.4.17}$$

The question is, what restriction is placed on the growth of the sequence $\{x_n\}$ by (1.4.17)?

By analogy with the case of difference *equations* with constant coefficients, the thing to try here is $x_n \leq K\alpha^n$. So suppose it is true that

$x_n \leq K\alpha^n$ for all $n = 0, 1, 2, \ldots, N$. Then from (1.4.17) with $n = N$ we find

$$x_{N+1} \leq K\alpha^N + K\alpha^{N-1} + N^2.$$

Let c be the positive real root of the equation $c^2 = c + 1$, and suppose that $\alpha > c$. Then $\alpha^2 > \alpha + 1$ and so $\alpha^2 - \alpha - 1 = t$, say, where $t > 0$. Hence

$$
\begin{aligned}
x_{N+1} &\leq K\alpha^{N-1}(1 + \alpha) + N^2 \\
&= K\alpha^{N-1}(\alpha^2 - t) + N^2 \qquad (1.4.18) \\
&= K\alpha^{N+1} - (tK\alpha^{N-1} - N^2).
\end{aligned}
$$

In order to insure that $x_{N+1} < K\alpha^{N+1}$ what we need is for $tK\alpha^{N-1} > N^2$. Hence as long as we choose

$$K > \max_{N \geq 2}\{N^2/t\alpha^{N-1}\}, \qquad (1.4.19)$$

in which the right member is clearly finite, the inductive step will go through.

The conclusion is that (1.4.17) implies that for every fixed $\epsilon > 0$, $x_n = O((c + \epsilon)^n)$, where $c = (1 + \sqrt{5})/2$. The same argument applies to the general situation that is expressed in

Theorem 1.4.1. *Let a sequence $\{x_n\}$ satisfy a recurrent inequality of the form*

$$x_{n+1} \leq b_0 x_n + b_1 x_{n-1} + \cdots + b_p x_{n-p} + G(n) \quad (n \geq p)$$

where $b_i \geq 0$ $(\forall i)$, $\sum b_i > 1$. Further, let c be the positive real root of * *the equation $c^{p+1} = b_0 c^p + \cdots + b_p$. Finally, suppose $G(n) = o(c^n)$. Then for every fixed $\epsilon > 0$ we have $x_n = O((c + \epsilon)^n)$.*

Proof: Fix $\epsilon > 0$, and let $\alpha = c + \epsilon$, where c is the root of the equation shown in the statement of the theorem. Since $\alpha > c$, if we let

$$t = \alpha^{p+1} - b_0 \alpha^p - \cdots - b_p$$

then $t > 0$. Finally, define

$$K = \max\left\{|x_0|, \frac{|x_1|}{\alpha}, \ldots, \frac{|x_p|}{\alpha^p}, \max_{n \geq p}\{\frac{G(n)}{t\alpha^{n-p}}\}\right\}.$$

---

* See exercise 10, below.

Then K is finite, and clearly $|x_j| \leq K\alpha^j$ for $j \leq p$. We claim that $|x_n| \leq K\alpha^n$ for all n, which will complete the proof.

Indeed, if the claim is true for $0, 1, 2, \ldots, n$, then

$$
\begin{aligned}
|x_{n+1}| &\leq b_0|x_0| + \cdots + b_p|x_{n-p}| + G(n) \\
&\leq b_0 K\alpha^n + \cdots + b_p K\alpha^{n-p} + G(n) \\
&= K\alpha^{n-p}\{b_0\alpha^p + \cdots + b_p\} + G(n) \\
&= K\alpha^{n-p}\{\alpha^{p+1} - t\} + G(n) \\
&= K\alpha^{n+1} - \{tK\alpha^{n-p} - G(n)\} \\
&\leq K\alpha^{n+1}.
\end{aligned}
$$

∎

Exercises for section 1.4

1. Solve the following recurrence relations
    (i) $x_{n+1} = x_n + 3$ $(n \geq 0; \; x_0 = 1)$
    (ii) $x_{n+1} = x_n/3 + 2$ $(n \geq 0; \; x_0 = 0)$
    (iii) $x_{n+1} = 2nx_n + 1$ $(n \geq 0; \; x_0 = 0)$
    (iv) $x_{n+1} = ((n+1)/n)x_n + n + 1$ $(n \geq 1; \; x_1 = 5)$
    (v) $x_{n+1} = x_n + x_{n-1}$ $(n \geq 1; \; x_0 = 0; \; x_1 = 3)$
    (vi) $x_{n+1} = 3x_n - 2x_{n-1}$ $(n \geq 1; \; x_0 = 1; \; x_1 = 3)$
    (vii) $x_{n+1} = 4x_n - 4x_{n-1}$ $(n \geq 1; \; x_0 = 1; \; x_1 = \xi)$
2. Find x_1 if the sequence $\mathbf{x}$ satisfies the Fibonacci recurrence relation and if furthermore $x_0 = 1$ and $x_n = o(1)$ $(n \to \infty)$.
3. Let x_n be the average number of trailing 0's in the binary expansions of all integers $0, 1, 2, \ldots, 2^n - 1$. Find a recurrence relation satisfied by the sequence $\{x_n\}$, solve it, and evaluate $\lim_{n \to \infty} x_n$.
4. For what values of a and b is it true that no matter what the initial values x_0, x_1 are, the solution of the recurrence relation $x_{n+1} = ax_n + bx_{n-1}$ $(n \geq 1)$ is guaranteed to be $o(1)$ $(n \to \infty)$?
5. Suppose $x_0 = 1$, $x_1 = 1$, and for all $n \geq 2$ it is true that $x_{n+1} \leq x_n + x_{n-1}$. Is it true that $\forall n : x_n \leq F_n$? Prove your answer.
6. Generalize the result of exercise 5, as follows. Suppose $x_0 = y_0$ and $x_1 = y_1$, and that $y_{n+1} = ay_n + by_{n-1}$ $(\forall n \geq 1)$. If furthermore, $x_{n+1} \leq$

$ax_n + bx_{n-1}$ $(\forall n \geq 1)$, can we conclude that $\forall n : x_n \leq y_n$? If not, describe conditions on a and b under which that conclusion would follow.

7. Find the asymptotic behavior in the form $x_n \sim$? $(n \to \infty)$ of the right side of (1.4.10).

8. Write out a complete proof of theorem 1.4.1.

9. Show by an example that the conclusion of theorem 1.4.1 may be false if the phrase 'for every fixed $\epsilon > 0\ldots$' were replaced by 'for every fixed $\epsilon \geq 0\ldots$.'

10. In theorem 1.4.1 we find the phrase '... the positive real root of ...' Prove that this phrase is justified, in that the equation shown always has exactly one positive real root. Exactly what special properties of that equation did you use in your proof?

1.5 Counting

For a given positive integer n, consider the set $\{1, 2, \ldots n\}$. We will denote this set by the symbol $[n]$, and we want to discuss the number of subsets of various kinds that it has. Here is a list of all of the subsets of $[2]$: $\emptyset$, $\{1\}$, $\{2\}$, $\{1, 2\}$. There are 4 of them.

We claim that the set $[n]$ has exactly 2^n subsets.

To see why, notice that we can construct the subsets of $[n]$ in the following way. Either choose, or don't choose, the element '1,' then either choose, or don't choose, the element '2,' etc., finally choosing, or not choosing, the element 'n.' Each of the n choices that you encountered could have been made in either of 2 ways. The totality of n choices, therefore, might have been made in 2^n different ways, so that is the number of subsets that a set of n objects has. ■

Next, suppose we have n distinct objects, and we want to arrange them in a sequence. In how many ways can we do that? For the first object in our sequence we may choose any one of the n objects. The second element of the sequence can be any of the remaining $n-1$ objects, so there are $n(n-1)$ possible ways to make the first two decisions. Then there are $n-2$ choices for the third element, and so we have $n(n-1)(n-2)$ ways to arrange the first three elements of the sequence. It is no doubt clear now that there are exactly $n(n-1)(n-2)\cdots 3 \cdot 2 \cdot 1 = n!$ ways to form the whole sequence.

Of the 2^n subsets of $[n]$, how many have exactly k objects in them?

The number of elements in a set is called its *cardinality*. The cardinality of a set S is denoted by $|S|$, so, for example, $|[6]| = 6$. A set whose cardinality is k is called a 'k-set,' and a subset of cardinality k is, naturally enough, a 'k-subset.' The question is, for how many subsets S of $[n]$ is it true that $|S| = k$?

We can construct k-subsets S of $[n]$ (written '$S \subseteq [n]$') as follows. Choose an element a_1 (n possible choices). Of the remaining $n-1$ elements, choose one ($n - 1$ possible choices), etc., until a sequence of k different elements has been chosen. Obviously there were $n(n - 1)(n - 2) \cdots (n - k + 1)$ ways in which we might have chosen that sequence, so the number of ways to choose an (ordered) sequence of k elements from $[n]$ is

$$n(n - 1)(n - 2) \cdots (n - k + 1) = n!/(n - k)!.$$

But there are more sequences of k elements than there are k-subsets, because any particular k-subset S will correspond to $k!$ different ordered sequences, namely all possible rearrangements of the elements of the subset. Hence the number of k-subsets of $[n]$ is equal to the number of k-sequences divided by $k!$. In other words, there are exactly $n!/k!(n - k)!$ k-subsets of a set of n objects.

The quantities $n!/k!(n - k)!$ are the famous *binomial coefficients*, and they are denoted by

$$\binom{n}{k} = \frac{n!}{k!(n - k)!} \quad (n \geq 0; 0 \leq k \leq n). \tag{1.5.1}$$

Some of their special values are

$$\binom{n}{0} = 1 \quad (\forall n \geq 0); \qquad \binom{n}{1} = n \quad (\forall n \geq 0);$$

$$\binom{n}{2} = n(n - 1)/2 \quad (\forall n \geq 0); \qquad \binom{n}{n} = 1 \quad (\forall n \geq 0).$$

It is convenient to define $\binom{n}{k}$ to be 0 if $k < 0$ or if $k > n$.

We can summarize the developments so far with

Theorem 1.5.1. *For each $n \geq 0$, a set of n objects has exactly 2^n subsets, and of these, exactly $\binom{n}{k}$ have cardinality k ($\forall k = 0, 1, \ldots, n$). There are*

35

exactly $n!$ *different sequences that can be formed from a set of n distinct objects.*

Since every subset of $[n]$ has *some* cardinality, it follows that

$$\sum_{k=0}^{n} \binom{n}{k} = 2^n \quad (n = 0, 1, 2, \ldots). \tag{1.5.2}$$

In view of the convention that we adopted, we might have written (1.5.2) as $\sum_k \binom{n}{k} = 2^n$, with no restriction on the range of the summation index k. It would then have been understood that the range of k is from $-\infty$ to ∞, and that the binomial coefficient $\binom{n}{k}$ vanishes unless $0 \le k \le n$.

In Table 1.5.1 we show the values of some of the binomial coefficients $\binom{n}{k}$. The rows of the table are thought of as labelled '$n = 0$,' '$n = 1$,' etc., and the entries within each row refer, successively, to $k = 0, 1, \ldots, n$. The table is called 'Pascal's triangle.'

```
                 1
               1   1
             1   2   1
           1   3   3   1
         1   4   6   4   1
       1   5  10  10   5   1
     1   6  15  20  15   6   1
   1   7  21  35  35  21   7   1
 1   8  28  56  70  56  28   8   1
```

....

Table 1.5.1: Pascal's triangle

Here are some facts about the binomial coefficients:

(a) Each row of Pascal's triangle is symmetric about the middle. That is,

$$\binom{n}{k} = \binom{n}{n-k} \quad (0 \le k \le n; n \ge 0).$$

(b) The sum of the entries in the nth row of Pascal's triangle is 2^n.

(c) Each entry is equal to the sum of the two entries that are immediately above it in the triangle.

The proof of (c) above can be interesting. What it says about the binomial coefficients is that

$$\binom{n}{k} = \binom{n-1}{k-1} + \binom{n-1}{k} \qquad ((n,k) \neq (0,0)). \qquad (1.5.3)$$

There are (at least) two ways to prove (1.5.3). The hammer-and-tongs approach would consist of expanding each of the three binomial coefficients that appears in (1.5.3), using the definition (1.5.1) in terms of factorials, and then cancelling common factors to complete the proof.

That would work (try it), but here's another way. Contemplate (this proof is by contemplation) the totality of k-subsets of $[n]$. The number of them is on the left side of (1.5.3). Sort them out into two piles: those k-subsets that contain '1' and those that don't. If a k-subset of $[n]$ contains '1,' then its remaining $k-1$ elements can be chosen in $\binom{n-1}{k-1}$ ways, and that accounts for the first term on the right of (1.5.3). If a k-subset does not contain '1,' then its k elements are all chosen from $[n-1]$, and that completes the proof of (1.5.3). ∎

The *binomial theorem* is the statement that $\forall n \geq 0$ we have

$$(1+x)^n = \sum_{k=0}^{n} \binom{n}{k} x^k. \qquad (1.5.4)$$

Proof: By induction on n. Eq. (1.5.4) is clearly true when $n = 0$, and if it is true for some n then multiply both sides of (1.5.4) by $(1+x)$ to obtain

$$(1+x)^{n+1} = \sum_k \binom{n}{k} x^k + \sum_k \binom{n}{k} x^{k+1}$$

$$= \sum_k \binom{n}{k} x^k + \sum_k \binom{n}{k-1} x^k$$

$$= \sum_k \{ \binom{n}{k} + \binom{n}{k-1} \} x^k$$

$$= \sum_k \binom{n+1}{k} x^k$$

which completes the proof. ∎

Now let's ask how big the binomial coefficients are, as an exercise in asymptotics. We will refer to the coefficients in row n of Pascal's triangle, that is, to

$$\binom{n}{0}, \binom{n}{1}, \ldots, \binom{n}{n}$$

as the coefficients of *order n*. Then, by (1.5.2) (or by (1.5.4) with $x = 1$), the sum of all of the coefficients of order n is 2^n. It is also fairly apparent, from an inspection of Table 1.5.1, that the largest one(s) of the coefficients of order n is (are) the one(s) in the middle.

More precisely, if n is odd, then the largest coefficients of order n are $\binom{n}{(n-1)/2}$ and $\binom{n}{(n+1)/2}$, whereas if n is even, the largest one is uniquely $\binom{n}{n/2}$.

It will be important, in some of the applications to algorithms later on in this book, for us to be able to pick out the largest term in a sequence of this kind, so let's see how we could *prove* that the biggest coefficients are the ones cited above.

For n fixed, we will compute the ratio of the $(k+1)^{st}$ coefficient of order n to the kth. We will see then that the ratio is larger than 1 if $k < (n-1)/2$ and is < 1 if $k > (n-1)/2$. That, of course, will imply that the $(k+1)^{st}$ coefficient is bigger than the kth, for such k, and therefore that the biggest one(s) must be in the middle.

The ratio is

$$\frac{\binom{n}{k+1}}{\binom{n}{k}} = \frac{n!/\{(k+1)!(n-k-1)!\}}{n!/\{k!(n-k)!\}}$$

$$= \frac{k!(n-k)!}{(k+1)!(n-k-1)!}$$

$$= (n-k)/(k+1)$$

and is > 1 if $k < (n-1)/2$, as claimed.

OK, the biggest coefficients are in the middle, but how big are they? Let's suppose that n is even, just to keep things simple. Then the biggest binomial coefficient of order n is

$$\binom{n}{n/2} = \frac{n!}{(n/2)!^2}$$

$$\sim \frac{(\frac{n}{e})^n \sqrt{2n\pi}}{\{(\frac{n}{2e})^{\frac{n}{2}} \sqrt{n\pi}\}^2} \tag{1.5.5}$$

$$= \sqrt{\frac{2}{n\pi}} 2^n$$

where we have used Stirling's formula (1.1.10).

Equation (1.5.5) shows that the single biggest binomial coefficient accounts for a very healthy fraction of the sum of *all* of the coefficients of order n. Indeed, the sum of all of them is 2^n, and the biggest one is $\sim \sqrt{2/n\pi} 2^n$. When n is large, therefore, the largest coefficient contributes a fraction $\sim \sqrt{2/n\pi}$ of the total.

If we think in terms of the subsets that these coefficients count, what we will see is that a large fraction of all of the subsets of an n-set have cardinality $n/2$, in fact $\Theta(n^{-.5})$ of them do. This kind of probabilistic thinking can be very useful in the design and analysis of algorithms. If we are designing an algorithm that deals with subsets of $[n]$, for instance, we should recognize that a large percentage of the customers for that algorithm will have cardinalities near $n/2$, and make every effort to see that the algorithm is fast for such subsets, even at the expense of possibly slowing it down on subsets whose cardinalities are very small or very large.

Exercises for section 1.5

1. How many subsets of even cardinality does $[n]$ have?

2. By observing that $(1+x)^a(1+x)^b = (1+x)^{a+b}$, prove that the sum of the squares of all binomial coefficients of order n is $\binom{2n}{n}$.

3. Evaluate the following sums in simple form.
   (i) $\sum_{j=0}^n j\binom{n}{j}$
   (ii) $\sum_{j=3}^n \binom{n}{j} 5^j$
   (iii) $\sum_{j=0}^n (j+1)3^{j+1}$

4. Find, by direct application of Taylor's theorem, the power series expansion of $f(x) = 1/(1-x)^{m+1}$ about the origin. Express the coefficients as certain binomial coefficients.

5. Complete the following twiddles.
   (i) $\binom{2n}{n} \sim ?$
   (ii) $\binom{n}{\lfloor \log_2 n \rfloor} \sim ?$
   (iii) $\binom{n}{\lfloor \theta n \rfloor} \sim ?$
   (iv) $\binom{n^2}{n} \sim ?$

6. How many ordered pairs of unequal elements of $[n]$ are there?

7. Which one of the numbers $\{2^j\binom{n}{j}\}_{j=0}^n$ is the biggest?

1.6 Graphs

A graph is a collection of *vertices*, certain unordered pairs of which are called its *edges*. To describe a particular graph we first say what its vertices are, and then we say which pairs of vertices are its edges. The set of vertices of a graph G is denoted by $V(G)$, and its set of edges is $E(G)$.

If v and w are vertices of a graph G, and if (v, w) is an edge of G, then we say that vertices v, w are *adjacent* in G.

Consider the graph G whose vertex set is $\{1, 2, 3, 4, 5\}$ and whose edges are the set of pairs (1,2), (2,3), (3,4), (4,5), (1,5). This is a graph of 5 vertices and 5 edges. A nice way to present a graph to an audience is to draw a picture of it, instead of just listing the pairs of vertices that are its edges. To draw a picture of a graph we would first make a point for each vertex, and then we would draw an arc between two vertices v and w if and only if (v, w) is an edge of the graph that we are talking about. The graph G of 5 vertices and 5 edges that we listed above can be drawn as shown in Fig. 1.6.1(a). It could also be drawn as shown in Fig. 1.6.1(b). They're both the same graph. Only the pictures are different, but the pictures aren't 'really' the graph; the graph is the vertex list and the edge list. The pictures are helpful to us in visualizing and remembering the graph, but that's all.

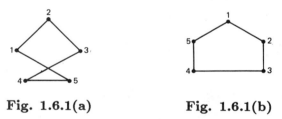

Fig. 1.6.1(a)   **Fig. 1.6.1(b)**

The number of edges that contain ('are incident with') a particular vertex v of a graph G is called the *degree* of that vertex, and is usually denoted by $\rho(v)$. If we add up the degrees of every vertex v of G we will have counted exactly two contributions from each edge of G, one at each of its endpoints. Hence, for every graph G we have

$$\sum_{v \in V(G)} \rho(v) = 2|E(G)|. \qquad (1.6.1)$$

Since the right-hand side is an even number, there must be an even number of odd numbers on the left side of (1.6.1). We have therefore proved that

every graph has an even number of vertices whose degrees are odd.* In Fig.
1.6.1 the degrees of the vertices are $\{2, 2, 2, 2, 2\}$ and the sum of the degrees
is $10 = 2|E(G)|$.

Next we're going to define a number of concepts of graph theory that
will be needed in later chapters. A fairly large number of terms will now be
defined, in rather a brief space. Don't try to absorb them all now, but read
through them and look them over again when the concepts are actually
used, in the sequel.

A *path* $\mathcal{P}$ in a graph G is a walk from one vertex of G to another,
where at each step the walk uses an edge of the graph. More formally,
it is a sequence $\{v_1, v_2, \ldots, v_k\}$ of vertices of G such that $\forall i = 1, k - 1 :$
$(v_i, v_{i+1}) \in E(G)$.

A graph is *connected* if there is a path between every pair of its vertices.

A path $\mathcal{P}$ is *simple* if its vertices are all distinct, *Hamiltonian* if it is
simple and visits every vertex of G exactly once, *Eulerian* if it uses every
edge of G exactly once.

A *subgraph* of a graph G is a subset S of its vertices to gether with a
subset of just those edges of G both of whose endpoints lie in S. An *induced
subgraph* of G is a subset S of the vertices of G together with *all* edges of
G both of whose endpoints lie in S. We would then speak of 'the subgraph
induced by S.'

In a graph G we can define an equivalence relation on the vertices as
follows. Say that v and w are equivalent if there is a path of G that joins
them. Let S be one of the equivalence classes of vertices of G under this
relation. The subgraph of G that S induces is called a *connected component*
of the graph G. A graph is *connected* if and only if it has exactly one
connected component.

A *cycle* is a closed path, *i.e.*, one in which $v_k = v_1$. A cycle is a *circuit*
if v_1 is the only repeated vertex in it. We may say that a circuit is a simple
cycle. We speak of Hamiltonian and Eulerian circuits of G as circuits of G
that visit, respectively, every vertex, or every edge, of a graph G.

Not every graph has a Hamiltonian path. The graph in Fig. 1.6.2(a)
has one and the graph in Fig. 1.6.2(b) doesn't.

---

* Did you realize that the number of people who shook hands an odd
number of times yesterday is an even number of people?

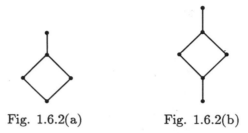

Fig. 1.6.2(a)   Fig. 1.6.2(b)

Likewise, not every graph has an Eulerian path. The graph in Fig. 1.6.3(a) has one and the graph in Fig. 1.6.3(b) doesn't.

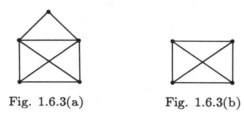

Fig. 1.6.3(a)   Fig. 1.6.3(b)

There is a world of difference between Eulerian and Hamiltonian paths, however. If a graph G is given, then thanks to the following elegant theorem of Euler, it is quite easy to decide whether or not G has an Eulerian path. In fact, the theorem applies also to *multigraphs*, which are graphs except that they are allowed to have several different edges joining the same pair of vertices.

Theorem 1.6.1. *A (multi-)graph has an Eulerian circuit (resp. path) if and only if it is connected and has no (resp. has exactly two) vertices of odd degree.*

Proof: Let G be a connected multigraph in which every vertex has even degree. We will find an Eulerian circuit in G. The proof for Eulerian paths will be similar, and is omitted.

The proof is by induction on the number of edges of G, and the result is clearly true if G has just one edge.

Hence suppose the theorem is true for all such multigraphs of fewer than m edges, and let G have m edges. We will construct an Eulerian circuit of G.

Begin at some vertex v and walk along some edge to a vertex w. Generically, having arrived at a vertex u, depart from u along an edge that hasn't been used yet, arriving at a new vertex, etc. The process halts when we arrive for the first time at a vertex v' such that all edges incident with v' have previously been walked on, so there is no exit.

We claim that $v' = v$, *i.e.*, we're back where we started. Indeed, if not, then we arrived at v' one more time than we departed from it, each time using a new edge, and finding no edges remaining at the end. Thus there were an odd number of edges of G incident with v', a contradiction.

Hence we are indeed back at our starting point when the walk terminates. Let W denote the sequence of edges along which we have so far walked. If W includes all edges of G then we have found an Euler tour and we are finished.

Else there are edges of G that are not in W. Erase all edges of W from G, thereby obtaining a (possibly disconnected multi-) graph G'. Let $C_1, \ldots, C_k$ denote the connected components of G'. Each of them has only vertices of even degree because that was true of G and of the walk W that we subtracted from G. Since each of the C_i has fewer edges than G had, there is, by induction, an Eulerian circuit in each of the connected components of G'.

We will thread them all together to make such a circuit for G itself.

Begin at the same v and walk along 0 or more edges of W until you arrive for the first time at a vertex q of component C_1. This will certainly happen because G is connected. Then follow the Euler tour of the edges of C_1, which will return you to vertex q. Then continue your momentarily interrupted walk W until you reach for the first time a vertex of C_2, which will surely happen because G is connected, etc., and the proof is complete. ∎

It is extremely difficult computationally to decide if a given graph has a Hamilton path or circuit. We will see in Chapter 5 that this question is typical of a breed of problems that are the main subject of that chapter, and are perhaps the most (in-)famous unsolved problems in theoretical computer science. Thanks to Euler's theorem (theorem 1.6.1) it is *easy* to decide if a graph has an *Eulerian* path or circuit.

Next we'd like to discuss graph *coloring*, surely one of the prettier parts of graph theory. Suppose that there are K colors available to us, and that we are presented with a graph G. A *proper* coloring of the vertices of G is an assignment of a color to each vertex of G in such a way that $\forall e \in E(G)$

the colors of the two endpoints of e are different. Fig. 1.6.4(a) shows a graph G and an attempt to color its vertices properly in 3 colors ('R,' 'Y' and 'B'). The attempt failed because one of the edges of G has had the same color assigned to both of its endpoints. In Fig. 1.6.4(b) we show the same graph with a successful proper coloring of its vertices in 4 colors.

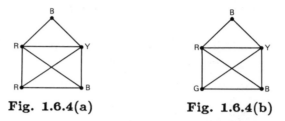

Fig. 1.6.4(a)   **Fig. 1.6.4(b)**

The *chromatic number* $\chi(G)$ of a graph G is the minimum number of colors that can be used in a proper coloring of the vertices of G. A *bipartite* graph is a graph whose chromatic number is ≤ 2, *i.e.*, it is a graph that can be 2-colored. That means that the vertices of a bipartite graph can be divided into two classes 'R' and 'Y' such that no edge of the graph runs between two 'R' vertices or between two 'Y' vertices. Bipartite graphs are most often drawn, as in Fig. 1.6.5, in two layers, with all edges running between layers.

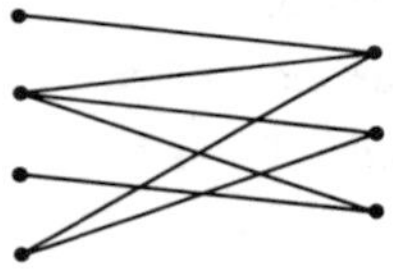

Fig. 1.6.5: A bipartite graph

The *complement* $\overline{G}$ of a graph G is the graph that has the same vertex set that G has and has an edge exactly where G does not have its edges. Formally,

$$E(\overline{G}) = \{(v, w) \mid v, w \in V(G); v \neq w; (v, w) \notin E(G)\}.$$

Here are some special families of graphs that occur so often that they rate special names. The *complete graph* K_n is the graph of n vertices in which every possible one of the $\binom{n}{2}$ edges is actually present. Thus K_2 is a single edge, K_3 looks like a triangle, etc. The *empty graph* $\overline{K}_n$ consists of n isolated vertices, *i.e.*, has no edges at all.

The *complete bipartite graph* $K_{m,n}$ has a set S of m vertices and a set T of n vertices. Its edge set is $E(K_{m,n}) = S \times T$. It has $|E(K_{m,n})| = mn$ edges. The n-cycle, C_n, is a graph of n vertices that are connected to form a single cycle. A *tree* is a graph that (a) is connected and (b) has no cycles. A tree is shown in Fig. 1.6.6.

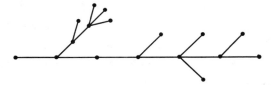

Fig. 1.6.6: A tree

It is not hard to prove that the following are equivalent descriptions of a tree.

(a) A tree is a graph that is connected and has no cycles.

(b) A tree is a graph G that is connected and for which $|E(G)| = |V(G)| - 1$.

(c) A tree is a graph G with the property that between every pair of distinct vertices there is a *unique* path.

If G is a graph and $S \subseteq V(G)$, then S is an *independent set* of vertices of G if no two of the vertices in S are adjacent in G. An independent set S is *maximal* if it is not a proper subset of another independent set of vertices of G. Dually, if a vertex subset S induces a complete graph, then we speak of a *complete subgraph* of G. A maximal complete subgraph of G is called a *clique*.

A graph might be *labeled* or *unlabeled*. The vertices of a labeled graph are numbered $1, 2, \ldots, n$. One difference that this makes is that there are a lot more labeled graphs than there are unlabeled graphs. There are, for example, 3 labeled graphs that have 3 vertices and 1 edge. They are shown in Fig. 1.6.7.

Fig. 1.6.7: Three labeled graphs...

There is, however, only 1 unlabeled graph that has 3 vertices and 1 edge, as shown in Fig. 1.6.8.

Fig. 1.6.8: ... but only one unlabeled graph

Most counting problems on graphs are much easier for labeled than for unlabeled graphs. Consider the following question: how many graphs are there that have exactly n vertices? Suppose first that we mean *labeled* graphs. A graph of n vertices has a maximum of $\binom{n}{2}$ edges. To construct a graph we would decide which of these possible edges would be used. We can make each of these $\binom{n}{2}$ decisions independently, and for every way of deciding where to put the edges we would get a different graph. Therefore the number of labeled graphs of n vertices is $2^{\binom{n}{2}} = 2^{n(n-1)/2}$.

If we were to ask the corresponding question for unlabeled graphs we would find it to be very hard. The answer is known, but the derivation involves Burnside's lemma about the action of a group on a set, and some fairly delicate counting arguments. We will state the approximate answer to this question, which is easy to write out, rather than the exact answer, which is not. If g_n is the number of unlabeled graphs of n vertices then

$$g_n \sim 2^{\binom{n}{2}}/n!.$$

Exercises for section 1.6

1. Show that a tree is a bipartite graph.

2. Find the chromatic number of the n-cycle.

3. Describe how you would find out, on a computer, if a given graph G is bipartite.

4. Given a positive integer K. Find two different graphs each of whose chromatic numbers is K.

5. Exactly how many labeled graphs of n vertices and E edges are there?

6. In how many labeled graphs of n vertices do vertices $\{1, 2, 3\}$ form an independent set?

7. How many cliques does an n-cycle have?

8. True or false: a **Hamiltonian** circuit is an induced cycle in a graph.

9. Which graph of n vertices has the largest number of independent sets? How many does it have?

10. Draw all of the connected, unlabeled graphs of 4 vertices.

11. Let G be a bipartite graph that has q connected components. Show that there are exactly 2^q ways to properly color the vertices of G in 2 colors.

12. Find a graph G of n vertices, other than the complete graph, whose chromatic number is equal to 1 plus the maximum degree of any vertex of G.

13. Let n be a multiple of 3. Consider a labeled graph G that consists of $n/3$ connected components, each of them a K_3. How many maximal independent sets does G have?

14. Describe the complement of the graph G in exercise 13 above. How many cliques does it have?

15. In how many labeled graphs of n vertices is the subgraph that is induced by vertices $\{1, 2, 3\}$ a triangle?

16. Let H be a labeled graph of L vertices. In how many labeled graphs of n vertices is the subgraph that is induced by vertices $\{1, 2, \ldots, L\}$ equal to H?

17. Devise an algorithm that will decide if a given graph, of n vertices and m edges, does or does not contain a triangle, in time $O(max(n^2, mn))$.

18. Prove that the number of labeled graphs of n vertices all of whose vertices have *even* degree is equal to the number of all labeled graphs of $n - 1$ vertices.

Chapter 2: Recursive Algorithms

2.1 Introduction

Here are two different ways to define $n!$, if n is a positive integer. The first definition is nonrecursive, the second is recursive.

(1) '$n!$ is the product of all of the whole numbers from 1 to n, inclusive.'

(2) 'If $n = 1$ then $n! = 1$, else $n! = n \cdot (n-1)!$.'

Let's concentrate on the second definition. At a glance, it seems illegal, because we're defining something, and in the definition the same 'something' appears. Another glance, however, reveals that the value of $n!$ is defined in terms of the value of the same function at a *smaller* value of its argument, *viz.* $n-1$. So we're really only using mathematical induction in order to validate the assertion that a function has indeed been defined for all positive integers n.

What is the practical import of the above? It's monumental. Many modern high-level computer languages can handle recursive constructs directly, and when this is so, the programmer's job may be considerably simplified. Among recursive languages are Pascal, PL/C, Lisp, APL, C, and many others. Programmers who use these languages should be aware of the power and versatility of recursive methods (conversely, people who like recursive methods should learn one of those languages!).

A formal 'function' module that would calculate $n!$ *nonrecursively* might look like this.

```
function fact(n);
{computes n! for given n > 0}
    fact := 1;
    for i := 1 to n do fact := i · fact;
end.
```

On the other hand a *recursive* $n!$ module is as follows.

```
function fact(n);
    if n = 1 then fact := 1 else fact := n · fact(n − 1);
end.
```

The hallmark of a recursive procedure is that it *calls itself*, with arguments that are in some sense smaller than before. Notice that there are no visible loops in the recursive routine. Of course there will be loops in the compiled machine-language program, so in effect the programmer is shifting many of the bookkeeping problems to the compiler (but *it* doesn't mind!).

Another advantage of recursiveness is that the *thought* processes are helpful. Mathematicians have known for years that induction is a marvellous method for proving theorems, making constructions, etc. Now computer scientists and programmers can profitably think recursively too, because recursive compilers allow them to express such thoughts in a natural way, and as a result many methods of great power are being formulated recursively, methods which, in many cases, might not have been developed if recursion were not readily available as a practical programming tool.

Observe next that the 'trivial case,' where $n = 1$, is handled separately, in the recursive form of the $n!$ program above. This trivial case is in fact essential, because it's the only thing that stops the execution of the program. In effect, the computer will be caught in a loop, reducing n by 1, until it reaches 1, then it will actually know the value of the function *fact*, and after that it will be able to climb back up to the original input value of n.

The overall structure of a recursive routine will always be something like this:

```
procedure calculate(list of variables);
    if {trivialcase} then do {trivialthing}
                else do
    {call calculate(smaller values of the variables)};
    {maybe do a few more things}
end.
```

In this chapter we're going to work out a number of examples of recursive algorithms, of varying sophistication. We will see how the recursive structure helps us to analyze the running time, or complexity, of the algorithms. We will also find that there is a bit of art involved in choosing the list of variables that a recursive procedure operates on. Sometimes the first list we think of doesn't work because the recursive call seems to need more detailed information than we have provided for it. So we try a larger list, and then perhaps it works, or maybe we need a still larger list ..., but more of this later.

Exercises for section 2.1

1. Write a *recursive* routine that will find the digits of a given integer n in the base b. There should be no visible loops in your program.

2.2 Quicksort

Suppose that we are given an array $x[1], \ldots, x[n]$ of n numbers. We would like to rearrange these numbers as necessary so that they end up in nondecreasing order of size. This operation is called *sorting* the numbers.

For instance, if we are given $\{9, 4, 7, 2, 1\}$, then we want our program to output the sorted array $\{1, 2, 4, 7, 9\}$.

There are many methods of sorting, but we are going to concentrate on methods that rely on only two kinds of basic operations, called *comparisons* and *interchanges*. This means that our sorting routine is allowed to

  (a) pick up two numbers ('keys') from the array, compare them, and decide which is larger.

  (b) interchange the positions of two selected keys.

Here is an example of a rather primitive sorting algorithm:

  (i) find, by successive comparisons, the smallest key

  (ii) interchange it with the first key

  (iii) find the second smallest key

  (iv) interchange it with the second key, etc. etc.

Here is a more formal algorithm that does the job above.

```
procedure slowsort(X: array[1..n]);
{sorts a given array into nondecreasing order}
  for r := 1 to n − 1 do
      for j := r + 1 to n do
          if x[j] < x[r]  then swap(x[j], x[r])
  end.{slowsort}
```

If you are wondering why we called this method 'primitive,' 'slowsort,' and other pejorative names, the reason will be clearer after we look at its complexity.

What is the cost of sorting n numbers by this method? We will look at two ways to measure that cost. First let's choose our unit of cost to be one comparison of two numbers, and then we will choose a different unit of cost, namely one interchange of position ('swap') of two numbers.

How many paired comparisons does the algorithm make? Reference to *procedure slowsort* shows that it makes one comparison for each value of $j = r + 1, \ldots, n$ in the inner loop. This means that the total number of comparisons is

$$f_1(n) = \sum_{r=1}^{n-1} \sum_{j=r+1}^{n} 1$$
$$= \sum_{r=1}^{n-1} (n - r)$$
$$= (n - 1)n/2.$$

The number of comparisons is $\Theta(n^2)$, which is quite a lot of comparisons for a sorting method to do. Not only that, but the method does that many comparisons regardless of the input array, *i.e.* its best case and its worst case are equally bad.

The Quicksort* method, which is the main object of study in this section, does a *maximum* of cn^2 comparisons, but *on the average* it does far fewer, a mere $O(n \log n)$ comparisons. This economy is much appreciated by those who sort, because sorting applications can be immense and time consuming. One popular sorting application is in alphabetizing lists of names. It is easy to imagine that some of those lists are very long, and that the replacement of $\Theta(n^2)$ by an average of $O(n \log n)$ comparisons is very welcome. An insurance company that wants to alphabetize its list of 5,000,000 policyholders will gratefully notice the difference between $n^2 = 25,000,000,000,000$ comparisons and $n \log n = 77,124,740$ comparisons.

If we choose as our unit of complexity the number of swaps of position, then the running time may depend strongly on the input array. In the 'slowsort' method described above, some arrays will need no swaps at all while others might require the maximum number of $(n-1)n/2$ (which arrays

---

* C. A. R. Hoare, *Comp. J.*, 5 (1962), 10-15.

need that many swaps?). If we average over all $n!$ possible arrangements of the input data, assuming that the keys are distinct, then it is not hard to see that the average number of swaps that *slowsort* needs is $\Theta(n^2)$.

Now let's discuss Quicksort. In contrast to the sorting method above, the basic idea of Quicksort is sophisticated and powerful. Suppose we want to sort the following list:

$$26, 18, 4, 9, 37, 119, 220, 47, 74 \qquad (2.2.1)$$

The number 37 in the above list is in a very intriguing position. Every number that precedes it is smaller than it is and every number that follows it is larger than it is. What that means is that *after sorting the list, the 37 will be in the same place it now occupies, the numbers to its left will have been sorted but will still be on its left, and the numbers on its right will have been sorted but will still be on its right.*

If we are fortunate enough to be given an array that has a 'splitter,' like 37, then we can

(a) sort the numbers to the left of the splitter, and then

(b) sort the numbers to the right of the splitter.

Obviously we have here the germ of a recursive sorting routine.

The fly in the ointment is that most arrays don't have splitters, so we won't often be lucky enough to find the state of affairs that exists in (2.2.1). However, we can make our own splitters, with some extra work, and that is the idea of the Quicksort algorithm. Let's state a preliminary version of the recursive procedure as follows (look carefully for how the procedure handles the trivial case where $n=1$).

> procedure *quicksortprelim*(x : an array of n numbers);
> {sorts the array x into nondecreasing order}
>   if $n \geq 2$ then
>   permute the array elements so as to create a splitter;
>   let $x[i]$ be the splitter that was just created;
>   *quicksortprelim*(the subarray $x_1, \ldots, x_{i-1}$) in place;
>   *quicksortprelim*(the subarray $x_{i+1}, \ldots, x_n$) in place
>   end.{*quicksortprelim*}

This preliminary version won't run, though. It looks like a recursive routine. It seems to call itself twice in order to get its job done. But it doesn't. It calls something that's just slightly different from itself in order to get its job done, and that won't work.

Observe the exact purpose of Quicksort, as described above. We are given an array of length n, and we want to sort it, *all of it*. Now look at the two 'recursive calls,' which really aren't quite. The first one of them sorts the array to the left of x_i. That is indeed a recursive call, because we can just change the 'n' to '$i-1$' and call Quicksort. The second recursive call is the problem. It wants to sort a portion of the array that doesn't begin at the beginning of the array. The routine Quicksort as written so far doesn't have enough flexibility to do that. So we will have to give it some more parameters.

Instead of trying to sort *all* of the given array, we will write a routine that sorts only the portion of the given array x that extends from $x[left]$ to $x[right]$, inclusive, where $left$ and $right$ are input parameters. This leads us to the second version of the routine:

> procedure *qksort*(x:array; *left, right*:integer);
> {sorts the subarray $x[left], \ldots, x[right]$}
>   if $right - left \geq 1$ then
>   create a splitter for the subarray in the ith array position;
> $qksort(x, left, i-1)$;
> $qksort(x, i+1, right)$
>   end.{*qksort*}

Once we have qksort, of course, Quicksort is no problem: we call qksort with $left := 1$ and $right := n$.

The next item on the agenda is the little question of how to create a splitter in an array. Suppose we are working with a subarray

$$x[left], x[left+1], \ldots, x[right].$$

The first step is to choose one of the subarray elements (the element itself, not the *position* of the element) to be the splitter, and the second step is to make it happen. The choice of the splitter element in the Quicksort algorithm is done very simply: *at random*. We just choose, using our favorite random number generator, one of the entries of the given subarray, let's call it T, and declare it to be the splitter. To repeat the parenthetical comment above, T is the *value* of the array entry that was chosen, not its *position* in the array. Once the value is selected, the position will be what it has to be, namely to the right of all smaller entries, and to the left of all larger entries.

The reason for making the random choice will become clearer after the smoke of the complexity discussion has cleared, but briefly it's this: the analysis of the average case complexity is realtively easy if we use the random choice, so that's a plus, and there are no minuses.

Second, we have now chosen T to be the value around which the sub-array will be split. The entries of the subarray must be moved so as to make T the splitter. To do this, consider the following algorithm.*

procedure $split(\mathbf{x}, left, right, i)$
{chooses at random an entry T of the subarray
      $[x_{left}, x_{right}]$, and splits the subarray around T}
{the output integer i is the position of T in the
      output array: $x[i] = T$};

```
1    L := a random integer in [left, right];
2    swap(x[left], x[L]);
3    {now the splitter is first in the subarray}
4    T := x[left];
5    i := left;
6    for j := left + 1 to right do
     begin
7        if x[j] < T  then
         begin
8            i := i + 1
             swap(x[i], x[j])
         end;
     end
9    swap(x[left], x[i])
10   end.{split}
```

We will now prove the correctness of *split*.

---

* Attributed to Nico Lomuto by Jon Bentley, *CACM* 27 (April 1984).

Theorem 2.2.1. *Procedure split correctly splits the array* x *around the chosen value T.*

Proof: We claim that as the loop in lines 7 and 8 is repeatedly executed for $j := left + 1$ *to right*, the following three assertions will always be true just *after* each execution of lines 7, 8:

(a) $x[left] = T$ and

(b) $x[r] < T$ for all $left < r \leq i$ and

(c) $x[r] \geq T$ for all $i < r \leq j$

Fig. 2.2.1 illustrates the claim.

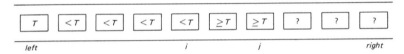

Fig. 2.2.1: Conditions (a), (b), (c)

To see this, observe first that (a), (b), (c) are surely true at the beginning, when $j = left + 1$. Next, if for some j they are true, then the execution of lines 7, 8 guarantee that they will be true for the next value of j.

Now look at (a), (b), (c) when $j = right$. It tells us that just prior to the execution of line 9 the condition of the array will be

(a) $x[left] = T$ and

(b) $x[r] < T$ for all $left < r \leq i$ and

(c) $x[r] \geq T$ for all $i < r \leq right$.

When line 9 executes, the array will be in the correctly split condition.

■

Now we can state a 'final' version of *qksort* (and therefore of Quicksort too).

> procedure *qksort*(x:array; *left*, *right*:integer);
> {sorts the subarray $x[left], \ldots, x[right]$};
> if $right - left \geq 1$ then
> $split(\text{x}, left, right, i)$;
> $qksort(\text{x}, left, i - 1)$;
> $qksort(\text{x}, i + 1, right)$
> end.{*qksort*}

> procedure *Quicksort*(x :array; *n*:integer)
> {sorts an array of length n};
> $qksort(\text{x}, 1, n)$
> end.{*Quicksort*}

Now let's consider the complexity of Quicksort. How long does it take to sort an array? Well, the amount of time will depend on exactly which array we happen to be sorting, and furthermore it will depend on how lucky we are with our random choices of splitting elements.

If we want to see Quicksort at its worst, suppose we have a really unlucky day, and that the random choice of the splitter element happens to be the smallest element in the array. Not only that, but suppose this kind of unlucky choice is repeated on each and every recursive call.

If the splitter element is the smallest array entry, then it won't do a whole lot of splitting. In fact, if the original array had n entries, then one of the two recursive calls will be to an array with no entries at all, and the other recursive call will be to an array of $n - 1$ entries. If $L(n)$ is the number of paired comparisons that are required in this extreme scenario, then, since the number of comparisons that are needed to carry out the call to *split* an array of length n is $n - 1$, it follows that

$$L(n) = L(n - 1) + n - 1 \qquad (n \geq 1; L(0) = 0).$$

Hence,

$$L(n) = (1 + 2 + \cdots + (n - 1)) = \Theta(n^2).$$

The worst-case behavior of Quicksort is therefore quadratic in n. In its worst moods, therefore, it is as bad as '*slowsort*' above.

Whereas the performance of *slowsort* is pretty much always quadratic, no matter what the input is, Quicksort is usually a lot faster than its worst case discussed above.

We want to show that *on the average* the running time of Quicksort is $O(n \log n)$.

The first step is to get quite clear about what the word 'average' refers to. We suppose that the entries of the input array **x** are all distinct. Then the performance of Quicksort can depend only on the sequence of size relationships in the input array and the choices of the random splitting elements.

The actual numerical values that appear in the input array are not in themselves important, except that, to simplify the discussion we will assume that they are all different. The only thing that will matter, then, will be the set of outcomes of all of the paired comparisons of two elements that are done by the algorithm. Therefore, we will assume, for the purposes of analysis, that the entries of the input array are exactly the set of numbers $1, 2, \ldots, n$ in some order.

There are $n!$ possible orders in which these elements might appear, so we are considering the action of Quicksort on just these $n!$ inputs.

Second, for each particular one of these inputs, the choices of the splitting elements will be made by choosing, at random, one of the entries of the array at each step of the recursion. We will also average over all such random choices of the splitting elements.

Therefore, when we speak of the function $F(n)$, the *average* complexity of Quicksort, we are speaking of the average number of *pairwise comparisons* of array entries that are made by Quicksort, where the averaging is done first of all over all $n!$ of the possible input orderings of the array elements, and second, for each such input ordering, we average also over all sequences of choices of the splitting elements.

Now let's consider the behavior of the function $F(n)$. What we are going to show is that $F(n) = O(n \log n)$.

The labor that $F(n)$ estimates has two components. First there are the pairwise comparisons involved in choosing a splitting element and rearranging the array about the chosen splitting value. Second there are the comparisons that are done in the two recursive calls that follow the creation of a splitter.

As we have seen, the number of comparisons involved in splitting the array is $n - 1$. Hence it remains to estimate the number of comparisons in the recursive calls.

For this purpose, suppose we have rearranged the array about the splitting element, and that it has turned out that the splitting entry now occupies the ith position in the array.

Our next remark is that each value of $i = 1, 2, \ldots, n$ is equally likely to occur. The reason for this is that we chose the splitter originally by choosing a random array entry. Since all orderings of the array entries are equally likely, the one that we happened to have chosen was just as likely to have been the largest entry as to have been the smallest, or the 17th-from-largest, or whatever.

Since each value of i is equally likely, each i has probability $1/n$ of being chosen as the residence of the splitter.

If the splitting element lives in the ith array position, the two recursive calls to Quicksort will be on two subarrays, one of which has length $i - 1$ and the other of which has length $n - i$. The average numbers of pairwise comparisons that are involved in such recursive calls are $F(i-1)$ and $F(n-i)$, respectively. It follows that our average complexity function F satisfies the relation

$$F(n) = n - 1 + \frac{1}{n} \sum_{i=1}^{n} \{F(i - 1) + F(n - i)\} \qquad (n \geq 1). \qquad (2.2.2)$$

together with the initial value $F(0) = 0$.

How can we find the solution of the recurrence relation (2.2.2)? First let's simplify it a little by noticing that

$$\sum_{i=1}^{n} \{F(n - i)\} = F(n - 1) + F(n - 2) + \cdots + F(0)$$

$$= \sum_{i=1}^{n} \{F(i - 1)\} \qquad (2.2.3)$$

and so (2.2.2) can be written as

$$F(n) = n - 1 + \frac{2}{n} \sum_{i=1}^{n} F(i - 1). \qquad (2.2.4)$$

We can simplify (2.2.4) a lot by getting rid of the summation sign. This next step may seem like a trick at first (and it is!), but it's a trick that is used in so many different ways that now we call it a 'method.' What we do is first to multiply (2.2.4) by n, to get

$$nF(n) = n(n-1) + 2\sum_{i=1}^{n} F(i-1). \qquad (2.2.5)$$

Next, in (2.2.5) we replace n by $n-1$, yielding

$$(n-1)F(n-1) = (n-1)(n-2) + 2\sum_{i=1}^{n-1} F(i-1). \qquad (2.2.6)$$

Finally we subtract (2.2.6) from (2.2.5), and the summation sign obligingly disappears, leaving behind just

$$nF(n) - (n-1)F(n-1) = n(n-1) - (n-1)(n-2) + 2F(n-1). \quad (2.2.7)$$

After some tidying up, (2.2.7) becomes

$$F(n) = (1 + \frac{1}{n})F(n-1) + (2 - \frac{2}{n}). \qquad (2.2.8)$$

which is exactly in the form of the general first-order recurrence relation that we discussed in section 1.4.

In section 1.4 we saw that to solve (2.2.8) the winning tactic is to change to a new variable y_n that is defined, in this case, by

$$F(n) = \frac{n+1}{n}\frac{n}{n-1}\frac{n-1}{n-2}\cdots\frac{2}{1}y_n \qquad (2.2.9)$$
$$= (n+1)y_n.$$

If we make the change of variable $F(n) = (n+1)y_n$ in (2.2.8), then it takes the form

$$y_n = y_{n-1} + 2(n-1)/n(n+1) \qquad (n \geq 1) \qquad (2.2.10)$$

as an equation for the y_n's ($y_0 = 0$).

The solution of (2.2.10) is obviously

$$y_n = 2\sum_{j=1}^{n} \frac{j-1}{j(j+1)}$$

$$= 2\sum_{j=1}^{n} \{\frac{2}{j+1} - \frac{1}{j}\}$$

$$= 2\sum_{j=1}^{n} \frac{1}{j} - 4n/(n+1).$$

Hence from (2.2.9),

$$F(n) = 2(n+1)\{\sum_{j=1}^{n} 1/j\} - 4n \qquad (2.2.11)$$

is the average number of pairwise comparisons that we do if we Quicksort an array of length n. Evidently $F(n) \sim 2n \log n$ $(n \to \infty)$ (see (1.1.7) with $g(t) = 1/t$), and we have proved

Theorem 2.2.2. *The average number of pairwise comparisons of array entries that Quicksort makes when it sorts arrays of n elements is exactly as shown in (2.2.11), and is $\sim 2n \log n$ $(n \to \infty)$.*

Quicksort is, on average, a very quick sorting method, even though its worst case requires a quadratic amount of labor.

Exercises for section 2.2

1. Write out an array of 10 numbers that contains no splitter. Write out an array of 10 numbers that contains 10 splitters.

2. Write a program that does the following. Given a positive integer n. Choose 100 random permutations of $[1, 2, \ldots, n]$,* and count how many of the 100 had at least one splitter. Execute your program for $n = 5, 6, \ldots, 12$ and tabulate the results.

3. Think of some method of sorting n numbers that isn't in the text. In the worst case, how many comparisons might your method do? How many swaps?

4. Consider the array

$$\mathbf{x} = \{2, 4, 1, 10, 5, 3, 9, 7, 8, 6\}$$

with *left* $= 1$ and *right* $= 10$. Suppose that the procedure *split* is called, and suppose the random integer L in step 1 happens to be 5. Carry out

---

* For a fast and easy way to do this see A. Nijenhuis and H. S. Wilf, *Combinatorial Algorithms*, 2nd ed. (New York: Academic Press, 1978), chap. 6.

the complete *split* algorithm (not on a computer; use pencil and paper). Particularly, record the condition of the array x after each value of j is processed in the *for j* = ... loop.

5. Suppose $H(0) = 1$ and $H(n) \leq 1 + \frac{1}{n}\sum_{i=1}^{n} H(i-1)$ $(n \geq 1)$. How big might $H(n)$ be?

6. If $Q(0) = 0$ and $Q(n) \leq n^2 + \sum_{i=1}^{n} Q(i-1)$ $(n \geq 1)$, how big might $Q(n)$ be?

7. (Research problem) Find the asymptotic behavior, for large n, of the probability that a randomly chosen permutation of n letters has a splitter.

2.3 Recursive graph algorithms

Algorithms on graphs are another rich area of applications of recursive thinking. Some of the problems are quite different from the ones that we have so far been studying in that they seem to need exponential amounts of computing time, rather than the polynomial times that were required for sorting problems.

We will illustrate the dramatically increased complexity with a recursive algorithm for the 'maximum independent set problem,' one which has received a great deal of attention in recent years.

Suppose a graph G is given. By an *independent set* of vertices of G we mean a set of vertices no two of which are connected by an edge of G. In the graph in Fig. 2.3.1 the set $\{1, 2, 6\}$ is an independent set and so is the set $\{1, 3\}$. The largest independent set of vertices in the graph shown there is the set $\{1, 2, 3, 6\}$. The problem of finding the size of the largest independent set in a given graph is computationally very difficult. All algorithms known to date require exponential amounts of time, in their worst cases, although no one has proved the nonexistence of fast (polynomial time) algorithms.

If the problem itself seems unusual, and maybe not deserving of a lot of attention, be advised that in Chapter 5 we will see that it is a member in good standing of a large family of very important computational problems (the 'NP-complete' problems) that are tightly bound together, in that if we can figure out better ways to compute any one of them, then we will be able to do all of them faster.

61

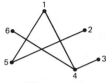

Fig. 2.3.1

Here is an algorithm for the independent set problem that is easy to understand and to program, although, of course, it may take a long time to run on a large graph G.

We are looking for the size of the largest independent set of vertices of G. Suppose we denote that number by $maxset(G)$. Fix some vertex of the graph, say vertex v^*. Let's distinguish two kinds of independent sets of vertices of G. There are those that contain vertex v^* and those that don't contain vertex v^*.

If an independent set S of vertices contains vertex v^*, then what does the rest of the set S consist of? The remaining vertices of S are an independent set in a smaller graph, namely the graph that is obtained from G by deleting vertex v^* as well as all vertices that are connected to vertex v^* by an edge. This latter set of vertices is called the *neighborhood* of vertex v^*, and is written $Nbhd(v^*)$.

The set S consists, therefore, of vertex v^* together with an independent set of vertices from the graph $G - \{v^*\} - Nbhd(v^*)$.

Now consider an independent set S that doesn't contain vertex v^*. In that case the set S is simply an independent set in the smaller graph $G - \{v^*\}$.

In either of the two cases above, the original problem has been reduced to a smaller one. Since the two cases are exhaustive of the possibilities, the original problem can always be done by solving two smaller ones.

We now have all of the ingredients of a recursive algorithm. Suppose we have found the two numbers $maxset(G - \{v^*\})$ and $maxset(G - \{v^*\} - Nbhd(v^*))$. Then, from the discussion above, we have the relation

$$maxset(G) = max\{maxset(G - \{v^*\}), 1 + maxset(G - \{v^*\} - Nbhd(v^*))\}.$$

We obtain the following recursive algorithm.

function $maxset1(G)$;
{returns the size of the largest independent set of
    vertices of G}
if G has no edges
    then $maxset1 := |V(G)|$
    else
        choose some nonisolated vertex v^* of G;
        $n_1 := maxset1(G - \{v^*\})$;
        $n_2 := maxset1(G - \{v^*\} - Nbhd(v^*))$;
        $maxset1 := max(n_1, 1 + n_2)$
end.{$maxset1$}

Example:

Here is an example of a graph G and the result of applying the $maxset1$ algorithm to it. Let the graph G be a 5-cycle. That is, it has 5 vertices and its edges are $(1,2), (2,3), (3,4), (4,5), (1,5)$. What are the two graphs on which the algorithm calls itself recursively?

Suppose we select vertex number 1 as the chosen vertex v in the algorithm. Then $G - \{1\}$ and $G - \{1\} - Nbhd(1)$ are respectively the two graphs shown in Fig. 2.3.2.

Fig. 2.3.2: $G - \{1\}$ $G - \{1\} - Nbhd(1)$

The reader should now check that the size of the largest independent set of G is equal to the larger of the two numbers $maxset1(G - \{1\})$, $1 + maxset1(G - \{1\} - Nbhd(1))$ in this example.

Of course the creation of these two graphs from the original input graph is just the beginning of the story, as far as the computation is concerned. Unbeknownst to the programmer, who innocently wrote the recursive routine $maxset1$ and then sat back to watch, the compiler will go ahead with the computation by generating a tree-full of graphs. In Fig. 2.3.3 we show

63

the collection of all of the graphs that the compiler might generate while executing a single call to *maxset*1 on the input graph of this example. In each case, the graph that is below and to the left of a given one is the one obtained by deleting a single vertex, and the one below and to the right of each graph is obtained by deleting a single vertex and its entire neighborhood.

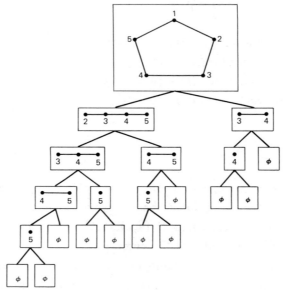

Fig. 2.3.3: A tree-full of graphs is created

Now we are going to study the complexity of *maxset*1. The results will be sufficiently depressing that we will then think about how to speed up the algorithm, and we will succeed in doing that to some extent.

To open the discussion, let's recall that in Chapter 0 it was pointed out that the complexity of a calculation is usefully expressed as a function of the number of bits of input data. In problems about graphs, however, it is more natural to think of the amount of labor as a function of n, the number of vertices of the graph. In problems about matrices it is more natural to use n, the size of the matrix, and so forth.

Do these distinctions alter the classification of problems into 'polynomial time do-able' vs. 'hard'? Take the graph problems, for instance. How many bits of input data does it take to describe a graph? Well, certainly we can march through the entire list of $n(n-1)/2$ pairs of vertices and check off the ones that are actually edges in the input graph to the problem. Hence we can describe a graph to a computer by making a list of $n(n-1)/2$ 0's and 1's. Each 1 represents a pair that is an edge, each 0 represents one that isn't an edge.

Thus $\Theta(n^2)$ bits describe a graph. Since n^2 is a polynomial in n, any function of the number of input data bits that can be bounded by a polynomial in same, can also be bounded by a polynomial in n itself. Hence, in the case of graph algorithms, the 'easiness' vs. 'hardness' judgment is not altered if we base the distinction on polynomials in n itself, rather than on polynomials in the number of bits of input data.

Hence, with a clear conscience, we are going to estimate the running time or complexity of graph algorithms in terms of the number of vertices of the graph that is input.

Now let's do this for algorithm *maxset*1 above.

The first step is to find out if G has any edges. To do this we simply have to look at the input data. In the worst case we might look at all of the input data, all $\Theta(n^2)$ bits of it. Then, if G actually has some edges, the additional labor needed to process G consists of two recursive calls on smaller graphs and one computation of the larger of two numbers.

If $F(G)$ denotes the total amount of computational labor that we do in order to find *maxset*1(G), then we see that

$$F(G) \leq cn^2 + F(G - \{v^*\}) + F(G - \{v^*\} - Nbhd(v^*)). \qquad (2.3.1)$$

Next, let $f(n) = \max_{|V(G)|=n} F(G)$, and take the maximum of (2.3.1) over all graphs G of n vertices. The result is that

$$f(n) \leq cn^2 + f(n-1) + f(n-2) \qquad (2.3.2)$$

because the graph $G - \{v^*\} - Nbhd(v^*)$ might have as many as $n-2$ vertices, and would have that many if v^* had exactly one neighbor.

Now it's time to 'solve' the recurrent inequality (2.3.2). Fortunately the hard work has all been done, and the answer is in theorem 1.4.1. That theorem was designed expressly for the analysis of recursive algorithms, and in this case it tells us that $f(n) = O((1.619^n))$. Indeed the number c in that theorem is $(1 + \sqrt{5})/2 = 1.61803....$ We chose the 'ϵ' that appears in the conclusion of the theorem simply by rounding c upwards.

What have we learned? Algorithm *maxset*1 will find the answer in a time of no more than $O(1.619^n)$ units if the input graph G has n vertices. This is a little improvement of the most simple-minded possible algorithm that one might think of for this problem, which is to examine every single subset of the vertices of G and ask if it is an independent set or not. That algorithm would take $\Theta(2^n)$ time units because there are 2^n subsets of vertices to look at. Hence we have traded in a 2^n for a 1.619^n by being a little bit cagey about the algorithm. Can we do still better?

There have in fact been a number of improvements of the basic *maxset*1 algorithm worked out. Of these the most successful is perhaps the one of Tarjan and Trojanowski that is cited in the bibliography at the end of this chapter, and later workers have slightly improved their results. We are not going to work out all of those ideas here, but instead we will show what kind of improvements on the basic idea will help us to do better in the time estimate.

We can obviously do better if we choose v^* in such a way as to be certain that it has at least *two* neighbors. If we were to do that then although we wouldn't affect the number of vertices of $G-\{v^*\}$ (always $n-1$) we would at least reduce the number of vertices of $G-\{v^*\}-Nbhd(v^*)$ as much as possible. So at least one of the two recursive calls of the program would be to a problem that is a little smaller than it was in *maxset*1.

So, as our next thought, we might replace the instruction 'choose some nonisolated vertex v^* of G' in *maxset*1 by an instruction 'choose some vertex v^* of G that has at least two neighbors.' Then we could be quite certain that $G-\{v^*\}-Nbhd(v^*)$ would have at most $n-3$ vertices.

What if there isn't any such vertex in the graph G? Then G would contain only vertices with 0 or 1 neighbors. The reader should have no difficulty in showing that such a graph G must be a collection of E disjoint edges together with a number m of isolated vertices.

The size of the largest independent set of vertices in such a graph is easy to find. A maximum independent set contains one vertex from each of the E edges and it contains all m of the isolated vertices. Hence in this case, $maxset = E + m = |V(G)| - |E(G)|$, and we obtain a second try at a good algorithm in the following form.

> procedure *maxset2(G)*;
> {returns the size of the largest independent set of
> vertices of *G*}
> if *G* has no vertex of degree ≥ 2
> then *maxset2* := $|V(G)| - |E(G)|$
> else
> choose a vertex v^* of degree ≥ 2;
> n_1 := *maxset2*($G - \{v^*\}$);
> n_2 := *maxset2*($G - \{v^*\} - Nbhd(v^*)$);
> *maxset2* := *max*($n_1, 1 + n_2$)
> end.{*maxset2*}

How much have we improved the complexity estimate? If we apply to *maxset2* the reasoning that led to (2.3.2) we find

$$f(n) \leq cn^2 + f(n-1) + f(n-3) \qquad (f(0) = 0; \; n = 2, 3, \ldots), \qquad (2.3.3)$$

where $f(n)$ is once more the worst-case time bound for graphs of n vertices.

Just as before, (2.3.3) is a recurrent inequality of the form that was studied at the end of section 1.4, in theorem 1.4.1. Using the conclusion of that theorem, we find from (2.3.3) that $f(n) = O((c + \epsilon)^n)$ where $c = 1.46557..$ is the positive root of the equation $c^3 = c^2 + 1$.

The net result of our effort to improve *maxset1* to *maxset2* has been to reduce the running-time bound from $O(1.619^n)$ to $O(1.47^n)$, which isn't a bad day's work. In the exercises below we will develop *maxset3*, whose running time will be $O(1.39^n)$. The idea will be that since in *maxset2* we were able to insure that v^* had at least two neighbors, why not try to insure that v^* has at least 3 of them?

As long as we have been able to reduce the time bound more and more by insuring that the selected vertex has lots of neighbors, why don't we keep it up, and insist that v^* should have 4 or more neighbors? Regrettably the method runs out of steam precisely at that moment. To see why, ask what the 'trivial case' would then look like. We would be working on a graph *G* in which no vertex has more than 3 neighbors. Well, what 'trivialthing' shall we do, in this 'trivial case'?

The fact is that there isn't any way of finding the maximum independent set in a graph where all vertices have ≤ 3 neighbors that's any faster than the general methods that we've already discussed. In fact, if one could find a fast method for that restricted problem it would have extremely important consequences, because we would then be able to do all graphs rapidly, not just those special ones.

We will learn more about this phenomenon in Chapter 5, but for the moment let's leave just the observation that the general problem of *maxset* turns out to be no harder than the special case of *maxset* in which no vertex has more than 3 neighbors.

Aside from the complexity issue, the algorithm *maxset* has shown how recursive ideas can be used to transform questions about graphs to questions about smaller graphs.

Here's another example of such a situation. Suppose G is a graph, and that we have a certain supply of colors available. To be exact, suppose we have K colors. We can then attempt to *color* the vertices of G properly in K colors (see section 1.6).

If we don't have enough colors, and G has lots of edges, this will not be possible. For example, suppose G is the graph of Fig. 2.3.4, and suppose we have just 3 colors available. Then there is no way to color the vertices without ever finding that both endpoints of some edge have the same color. On the other hand, if we have four colors available then we can do the job.

Fig. 2.3.4

There are many interesting computational and theoretical problems in the area of coloring of graphs. Just for its general interest, we are going to mention the four-color theorem, and then we will turn to a study of some of the computational aspects of graph coloring.

First, just for general cultural reasons, let's slow down for a while and discuss the relationship between graph colorings in general and the four-color problem, even though it isn't directly relevant to what we're doing.

The original question was this. Suppose that a delegation of Earthlings were to visit a distant planet and find there a society of human beings. Since that race is well known for its squabbling habits, you can be sure that the planet will have been carved up into millions of little countries, each with its own ruling class, system of government, etc., and of course, all at war with each other. The delegation wants to escape quickly, but before doing so it draws a careful map of the 5,000,000 countries into which the planet has been divided. To make the map easier to read, the countries are then colored in such a way that whenever two countries share a stretch of border they are of two different colors. Surprisingly, it was found that the coloring could be done using only red, blue, yellow and green.

It was noticed over 100 years ago that no matter how complicated a map is drawn, and no matter how many countries are involved, it seems to be possible to color the countries in such a way that

(a) every pair of countries that have a common stretch of border have different colors and

(b) no more than *four* colors are used in the entire map.

It was then conjectured that four colors are always sufficient for the proper coloring of the countries of any map at all. Settling this conjecture turned out to be a very hard problem. It was finally solved in 1976 by K. Appel and W. Haken* by means of an extraordinary proof with two main ingredients. First they showed how to reduce the general problem to only a *finite* number of cases, by a mathematical argument. Then, since the 'finite number' was over 1800, they settled all of those cases with quite a lengthy computer calculation. So now we have the 'Four Color Theorem,' which asserts that no matter how we carve up the plane or the sphere into countries, we will always be able to color those countries with at most four colors so that countries with a common frontier are colored differently.

We can change the *map* coloring problem into a *graph* coloring problem as follows. Given a map. From the map we will construct a graph G. There will be a vertex of G corresponding to each country on the map. Two of these vertices will be connected by an edge of the graph G if the two countries that they correspond to have a common stretch of border (we keep saying 'stretch of border' to emphasize that if the two countries have just a single point in common they are allowed to have the same color).

---

* Every planar map is four colorable, *Bull. Amer. Math. Soc.*, **82** (1976), 711-712.

As an illustration of this construction, we show in Fig. 2.3.5(a) a map of a distant planet, and in Fig. 2.3.5(b) the graph that results from the construction that we have just described.

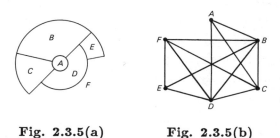

Fig. 2.3.5(a)   Fig. 2.3.5(b)

By a 'planar graph' we mean a graph G that can be drawn in the plane in such a way that two edges never cross (except that two edges at the same vertex have that vertex in common). The graph that results from changing a map of countries into a graph as described above is always a planar graph. In Fig. 2.3.6(a) we show a planar graph G. This graph doesn't look planar because two of its edges cross. However, that isn't the graph's fault, because with a little more care we might have drawn *the same graph* as in Fig. 2.3.6(b), in which its planarity is obvious. Don't blame the graph if it doesn't look planar. It might be planar anyway!

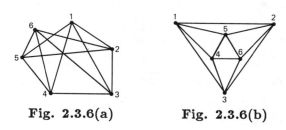

Fig. 2.3.6(a)   Fig. 2.3.6(b)

The question of recognizing whether a given graph is planar is itself a formidable problem, although the solution, due to J. Hopcroft and R. E. Tarjan,* is an algorithm that makes the decision in *linear time, i.e.* in $O(V)$ time for a graph of V vertices.

Although every *planar* graph can be properly colored in four colors, there are still all of those other graphs that are not planar to deal with.

---

* Efficient planarity testing, *J. Assoc. Comp. Mach.* **21** (1974), 549-568.

For any one of those graphs we can ask, if a positive integer K is given, whether or not its vertices can be K-colored properly.

As if that question weren't hard enough, we might ask for even more detail, namely about the *number* of ways of properly coloring the vertices of a graph. For instance, if we have K colors to work with, suppose G is the *empty* graph $\overline{K}_n$, that is, the graph of n vertices that has no edges at all. Then G has quite a large number of proper colorings, K^n of them, to be exact. Other graphs of n vertices have fewer proper colorings than that, and an interesting computational question is to count the proper colorings of a given graph.

We will now find a recursive algorithm that will answer this question. Again, the complexity of the algorithm will be exponential, but as a small consolation we note that no polynomial time algorithm for this problem is known.

Choose an edge e of the graph, and let its endpoints be v and w. Now delete the edge e from the graph, and let the resulting graph be called $G - \{e\}$. Then we will distinguish two kinds of proper colorings of $G - \{e\}$: those in which vertices v and w have the same color and those in which v and w have different colors. Obviously the number of proper colorings of $G - \{e\}$ in K colors is the sum of the numbers of colorings of each of these two kinds.

Consider the proper colorings in which vertices v and w have the same color. We claim that the number of such colorings is equal to the number of *all* colorings of a certain new graph $G/\{e\}$, whose construction we now describe:

The vertices of $G/\{e\}$ consist of the vertices of G other than v or w and one new vertex that we will call 'vw' (so $G/\{e\}$ will have one less vertex than G has).

Now we describe the *edges* of $G/\{e\}$. First, if a and b are two vertices of $G/\{e\}$ neither of which is the new vertex 'vw', then (a, b) is an edge of $G/\{e\}$ if and only if it is an edge of G. Second, (vw, b) is an edge of $G/\{e\}$ if and only if either (v, b) or (w, b) (or both) is an edge of G.

We can think of this as 'collapsing' the graph G by imagining that the edges of G are elastic bands, and that we squeeze vertices v and w together into a single vertex. The result is $G/\{e\}$ (anyway, it is if we replace any resulting double bands by single ones!).

71

In Fig. 2.3.7(a) we show a graph G of 7 vertices and a chosen edge e. The two endpoints of e are v and w. In Fig. 2.3.7(b) we show the graph $G/\{e\}$ that is the result of the construction that we have just described.

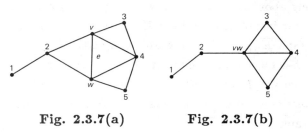

Fig. 2.3.7(a)   **Fig. 2.3.7(b)**

The point of the construction is the following

Lemma 2.3.1. *Let v and w be two vertices of G such that $e = (v, w) \in E(G)$. Then the number of proper K-colorings of $G - \{e\}$ in which v and w have the same color is equal to the number of all proper colorings of the graph $G/\{e\}$.*

Proof: Suppose $G/\{e\}$ has a proper K-coloring. Color the vertices of $G - \{e\}$ itself in K colors as follows. Every vertex of $G - \{e\}$ other than v or w keeps the same color that it has in the coloring of $G/\{e\}$. Vertex v and vertex w each receive the color that vertex vw has in the coloring of $G/\{e\}$. Now we have a K-coloring of the vertices of $G - \{e\}$.

It is a proper coloring because if f is any edge of $G - \{e\}$ then the two endpoints of f have different colors. Indeed, this is obviously true if neither endpoint of f is v or w because the coloring of $G/\{e\}$ was a proper one. There remains only the case where one endpoint of f is, say, v and the other one is some vertex x other than v or w. But then the colors of v and x must be different because vw and x were joined in $G/\{e\}$ by an edge, and therefore must have gotten different colors there. ∎

To get back to the main argument, we were trying to compute the number of proper K-colorings of $G - \{e\}$. We observed that in any K-coloring v and w have either the same or different colors. We have shown that the number of colorings in which they receive the same color is equal to the number of all proper colorings of a certain smaller (one less vertex) graph $G/\{e\}$. It remains to look at the case where vertices v and w receive different colors.

Lemma 2.3.2. *Let $e = (v, w)$ be an edge of G. Then the number of proper K-colorings of $G - \{e\}$ in which v and w have different colors is equal to the number of all proper K-colorings of G itself.*

Proof: Obvious (isn't it?). ■

Now let's put together the results of the two lemmas above. Let $P(K; G)$ denote the number of ways of properly coloring the vertices of a given graph G. Then lemmas 2.3.1 and 2.3.2 assert that

$$P(K; G - \{e\}) = P(K; G/\{e\}) + P(K; G)$$

or if we solve for $P(K; G)$, then we have

$$P(K; G) = P(K; G - \{e\}) - P(K; G/\{e\}). \qquad (2.3.4)$$

The quantity $P(K; G)$, the number of ways of properly coloring the vertices of a graph G in K colors, is called *the chromatic polynomial of G.*

We claim that it is, in fact, a polynomial in K of degree $|V(G)|$. For instance, if G is the complete graph of n vertices then obviously $P(K, G) = K(K-1) \cdots (K - n + 1)$, and that is indeed a polynomial in K of degree n.

Proof of claim: The claim is certainly true if G has just one vertex. Next suppose the assertion is true for graphs of $< V$ vertices, then we claim it is true for graphs of V vertices also. This is surely true if G has V vertices and no edges at all. Hence, suppose it is true for all graphs of V vertices and fewer than E edges, and let G have V vertices and E edges. Then (2.3.4) implies that $P(K; G)$ is a polynomial of the required degree V because $G - \{e\}$ has fewer edges than G does, so its chromatic polynomial is a polynomial of degree V. $G/\{e\}$ has fewer vertices than G has, and so $P(K; G/\{e\})$ is a polynomial of lower degree. The claim is proved, by induction. ■

Equation (2.3.4) gives a recursive algorithm for computing the chromatic polynomial of a graph G, since the two graphs that appear on the right are both 'smaller' than G, one in the sense that it has fewer edges than G has, and the other in that it has fewer vertices. The algorithm is the following.

```
function chrompoly(G:graph): polynomial;
{computes the chromatic polynomial of a graph G}
   if G has no edges  then chrompoly:=K|V(G)|
                   else
      choose an edge e of G;
      chrompoly:=chrompoly(G − {e})−chrompoly(G/{e})
   end.{chrompoly}
```

Next we are going to look at the complexity of the algorithm *chrompoly* (we will also refer to it as 'the delete-and-identify' algorithm). The graph *G* can be input in any one of a number of ways. For example, we might input the full list of edges of *G*, as a list of pairs of vertices.

The first step of the computation is to choose the edge *e* and to create the edge list of the graph *G* − {*e*}. The latter operation is trivial, since all we have to do is to ignore one edge in the list.

Next we call *chrompoly* on the graph *G* − {*e*}.

The third step is to create the edge list of the collapsed graph *G*/{*e*} from the edge list of *G* itself. That involves some work, but it is rather routine, and its cost is linear in the number of edges of *G*, say $c|E(G)|$.

Finally we call *chrompoly* on the graph *G*/{*e*}.

Let $F(V, E)$ denote the maximum cost of calling *chrompoly* on any graph of at most *V* vertices and at most *E* edges. Then we see at once that

$$F(V, E) \leq F(V, E-1) + cE + F(V-1, E-1) \qquad (2.3.5)$$

together with $F(V, 0) = 0$. If we put, successively, $E = 1, 2, 3$, we find that $F(V, 1) \leq c$, $F(V, 2) \leq 4c$, and $F(V, 3) \leq 11c$. Hence we seek a solution of (2.3.5) in the form $F(V, E) \leq f(E)c$, and we quickly find that if

$$f(E) = 2f(E-1) + E \qquad (f(0) = 0) \qquad (2.3.6)$$

then we will have such a solution.

Since (2.3.6) is a first-order difference equation of the form (1.4.5), we find that

$$f(E) = 2^E \sum_{j=0}^{E} j2^{-j} \qquad (2.3.7)$$
$$\sim 2^{E+1}.$$

The last '$\sim$' follows from the evaluation $\sum j 2^{-j} = 2$ that we discussed in section 1.3.

To summarize the developments so far, then, we have found out that the chromatic polynomial of a graph can be computed recursively by an algorithm whose cost is $O(2^E)$ for graphs of E edges. This is exponential cost, and such computations are prohibitively expensive except for graphs of very modest numbers of edges.

Of course the mere fact that our proved time estimate is $O(2^E)$ doesn't necessarily mean that the algorithm can be that slow, because maybe our complexity analysis wasn't as sharp as it might have been. However, consider the graph $G(s,t)$ that consists of s disjoint edges and t isolated vertices, for a total of $2s + t$ vertices altogether. If we choose an edge of $G(s,t)$ and delete it, we get $G(s-1, t+2)$, whereas the graph $G/\{e\}$ is $G(s-1, t+1)$. Each of these two new graphs has $s-1$ edges.

We might imagine arranging the computation so that the extra isolated vertices will be 'free,' *i.e.*, will not cost any additional labor. Then the work that we do on $G(s,t)$ will depend only on s, and will be twice as much as the work we do on $G(s-1, \cdot)$. Therefore $G(s,t)$ will cost at least 2^s operations, and our complexity estimate wasn't a mirage, there really are graphs that make the algorithm do an amount $2^{|E(G)|}$ of work.

Considering the above remarks it may be surprising that there is a slightly different approach to the complexity analysis that leads to a time bound (for the same algorithm) that is a bit sharper than $O(2^E)$ in many cases (the work of the complexity analyst is never finished!). Let's look at the algorithm *chrompoly* in another way.

For a graph G we can define a number $\gamma(G) = |V(G)| + |E(G)|$, which is rather an odd kind of thing to define, but it has a nice property with respect to this algorithm, namely that whatever G we begin with, we will find that

$$\gamma(G - \{e\}) = \gamma(G) - 1; \quad \gamma(G/\{e\}) \le \gamma(G) - 2. \qquad (2.3.8)$$

Indeed, if we delete the edge e then γ must drop by 1, and if we collapse the graph on the edge e then we will have lost one vertex and at least one edge, so γ will drop by at least 2.

Hence, if $h(\gamma)$ denotes the maximum amount of labor that *chrompoly* does on any graph G for which

$$|V(G)| + |E(G)| \leq \gamma \tag{2.3.9}$$

then we claim that

$$h(\gamma) \leq h(\gamma - 1) + h(\gamma - 2) \qquad (\gamma \geq 2). \tag{2.3.10}$$

Indeed, if G is a graph for which (2.3.9) holds, then if G has any edges at all we can do the delete-and-identify step to prove that the labor involved in computing the chromatic polynomial of G is at most the quantity on the right side of (2.3.10). Else, if G has no edges then the labor is 1 unit, which is again at most equal to the right side of (2.3.10), so the result (2.3.10) follows.

With the initial conditions $h(0) = h(1) = 1$ the solution of the recurrent inequality (2.3.10) is obviously the relation $h(\gamma) \leq F_\gamma$, where F_γ is the Fibonacci number. We have thereby proved that the time complexity of the algorithm *chrompoly* is

$$O(F_{|V(G)|+|E(G)|}) = O\left(\left(\frac{1 + \sqrt{5}}{2}\right)^{|V(G)|+|E(G)|}\right) \tag{2.3.11}$$
$$= O(1.62^{|V(G)|+|E(G)|}).$$

This analysis does not, of course, contradict the earlier estimate, but complements it. What we have shown is that the labor involved is always

$$O\left(min(2^{|E(G)|}, 1.62^{|V(G)|+|E(G)|})\right). \tag{2.3.12}$$

On a graph with 'few' edges relative to its number of vertices (how few?) the first quantity in the parentheses in (2.3.12) will be the smaller one, whereas if G has more edges, then the second term is the smaller one. In either case the overall judgment about the speed of the algorithm (it's slow!) remains.

Exercises for section 2.3

1. Let G be a cycle of n vertices. What is the size of the largest independent set of vertices in $V(G)$?

2. Let G be a path of n vertices. What is the size of the largest independent set of vertices in $V(G)$?

3. Let G be a connected graph in which every vertex has degree 2. What must such a graph consist of? Prove.

4. Let G be a connected graph in which every vertex has degree ≤ 2. What must such a graph look like?

5. Let G be a not-necessarily-connected graph in which every vertex has degree ≤ 2. What must such a graph look like? What is the size of the largest independent set of vertices in such a graph? How long would it take you to calculate that number for such a graph G? How would you do it?

6. Write out algorithm *maxset3*, which finds the size of the largest independent set of vertices in a graph. Its trivial case will occur if G has no vertex of degree ≥ 3. Otherwise, it will choose a vertex v^* of degree ≥ 3 and proceed as in *maxset2*.

7. Analyze the complexity of your algorithm *maxset3* from exercise 6 above.

8. Use (2.3.4) to prove by induction that $P(K; G)$ is a polynomial in K of degree $|V(G)|$. Then show that if G is a tree then $P(K; G) = K(K - 1)^{|V(G)|-1}$.

9. Write out an algorithm that will change the vertex adjacency matrix of a graph G to the vertex adjacency matrix of the graph $G/\{e\}$, where e is a given edge of G.

10. How many edges must G have before the second quantity inside the 'O' in (2.3.12) is the smaller of the two?

11. Let $\alpha(G)$ be the size of the largest independent set of vertices of a graph G, let $\chi(G)$ be its chromatic number, and let $n = |V(G)|$. Show that, for every G, $\alpha(G) \geq n/\chi(G)$.

2.4 Fast matrix multiplication

Everybody knows how to multiply two 2×2 matrices. If we want to calculate

$$\begin{pmatrix} c_{11} & c_{12} \\ c_{21} & c_{22} \end{pmatrix} = \begin{pmatrix} a_{11} & a_{12} \\ a_{21} & a_{22} \end{pmatrix} \begin{pmatrix} b_{11} & b_{12} \\ b_{21} & b_{22} \end{pmatrix} \tag{2.4.1}$$

then, 'of course,'

$$c_{i,j} = \sum_{k=1}^{2} a_{i,k} b_{k,j} \qquad (i, j = 1, 2). \tag{2.4.2}$$

Now look at (2.4.2) a little more closely. In order to calculate each one of the 4 $c_{i,j}$'s we have to do 2 multiplications of numbers. The cost of multiplying two 2×2 matrices is therefore 8 multiplications of numbers. If we measure the cost in units of additions of numbers, the cost is 4 such additions. Hence, the matrix multiplication method that is shown in (2.4.1) has a complexity of 8 multiplications of numbers and 4 additions of numbers.

This may seem rather unstartling, but the best ideas often have humble origins.

Suppose we could find another way of multiplying two 2×2 matrices in which the cost was only 7 multiplications of numbers, together with more than 4 additions of numbers. Would that be a cause for dancing in the streets, or would it be just a curiosity, of little importance? In fact, it would be extremely important, and the consequences of such a step were fully appreciated only in 1969 by V. Strassen, to whom the ideas that we are now discussing are due.*

What we're going to do next in this section is the following:

(a) describe another way of multiplying two 2×2 matrices in which the cost will be only 7 multiplications of numbers plus a bunch of additions of numbers, and

(b) convince you that it was worth the trouble.

The usefulness of the idea stems from the following amazing fact: if two 2×2 matrices can be multiplied with only 7 multiplications of numbers, then two $N \times N$ matrices can be multiplied using only $O(N^{2.81\cdots})$ multiplications of numbers instead of the N^3 such multiplications that the usual method involves (the number '2.81...' is $\log_2 7$).

In other words, if we can reduce the number of multiplications of numbers that are needed to multiply two 2×2 matrices, then that improvement will show up in the *exponent* of N when we measure the complexity of multiplying two $N \times N$ matrices. The reason, as we will see, is that the little improvement will be pyramided by numerous recursive calls to the 2×2 procedure– but we get ahead of the story.

Now let's write out another way to do the 2×2 matrix multiplication that is shown in (2.4.1). Instead of doing it *á là* (2.4.2), try the following

---

* V. Strassen, Gaussian elimination is not optimal, *Numerische Math.* **13** (1969), 354-6.

11-step approach.

First compute, from the input 2×2 matrices shown in (2.4.1), the following 7 quantities:

$$I = (a_{12} - a_{22}) \times (b_{21} + b_{22})$$
$$II = (a_{11} + a_{22}) \times (b_{11} + b_{22})$$
$$III = (a_{11} - a_{21}) \times (b_{11} + b_{12})$$
$$IV = (a_{11} + a_{12}) \times b_{22} \qquad (2.4.3)$$
$$V = a_{11} \times (b_{12} - b_{22})$$
$$VI = a_{22} \times (b_{21} - b_{11})$$
$$VII = (a_{21} + a_{22}) \times b_{11}$$

and then calculate the 4 entries of the product matrix $C = AB$ from the 4 formulas

$$c_{11} = I + II - IV + VI$$
$$c_{12} = IV + V$$
$$c_{21} = VI + VII \qquad (2.4.4)$$
$$c_{22} = II - III + V - VII.$$

The first thing to notice about this seemingly overelaborate method of multiplying 2×2 matrices is that only 7 multiplications of numbers are used (count the '$\times$' signs in (2.4.3)). 'Well yes,' you might reply, 'but 18 additions are needed, so where is the gain?'

It will turn out that multiplications are more important than additions, not because computers can do them faster, but because when the routine is called *recursively* each '$\times$' operation will turn into a multiplication of two big matrices whereas each '$\pm$' will turn into an addition or subtraction of two big matrices, and that's much cheaper.

Next we're going to describe how Strassen's method (equations (2.4.3), (2.4.4)) of multiplying 2×2 matrices can be used to speed up multiplications of $N \times N$ matrices. The basic idea is that we will partition each of the large matrices into four smaller ones and multiply them together using (2.4.3), (2.4.4).

Suppose that N is a power of 2, say $N = 2^n$, and let there be given two $N \times N$ matrices, A and B. We imagine that A and B have each been partitioned into four $2^{n-1} \times 2^{n-1}$ matrices, and that the product matrix

C is similarly partitioned. Hence we want to do the matrix multiplication that is indicated by

$$\begin{pmatrix} C_{11} & C_{12} \\ C_{21} & C_{22} \end{pmatrix} = \begin{pmatrix} A_{11} & A_{12} \\ A_{21} & A_{22} \end{pmatrix} \begin{pmatrix} B_{11} & B_{12} \\ B_{21} & B_{22} \end{pmatrix} \qquad (2.4.5)$$

where now each of the capital letters represents a $2^{n-1} \times 2^{n-1}$ matrix.

To do the job in (2.4.5) we use exactly the 11 formulas that are shown in (2.4.3) and (2.4.4), except that the lower-case letters are now all upper case. Suddenly we very much appreciate the reduction of the number of '$\times$' signs because it means one less multiplication of large matrices, and we don't so much mind that it has been replaced by 10 more '$\pm$' signs, at least not if N is very large.

This yields the following recursive procedure for multiplying large matrices.

> function $MatrProd(A, B$: matrix; N:integer):matrix;
> $\{MatrProd$ is AB, where A and B are $N \times N\}$
> $\{$uses Strassen method$\}$
> **if** N is not a power of 2 **then**
> > border A and B by rows and columns of 0's until
> > their size is the next power of 2 and change N;
>
> **if** $N = 1$ **then** $MatrProd := AB$
> > > **else**
> >
> > partition A and B as shown in (2.4.5);
> > $I := MatrProd(A_{11} - A_{22}, B_{21} + B_{22}, N/2)$;
> > $II := MatrProd(A_{11} + A_{22}, B_{11} + B_{22}, N/2)$;
> > > etc. etc., through all 11 of the formulas
> > > shown in (2.4.3), (2.4.4), ending with ...
> >
> > $C_{22} := II - III + V - VII$
>
> end.$\{MatrProd\}$

Note that this procedure calls itself recursively 7 times. The plus and minus signs in the program each represent an addition or subtraction of two *matrices*, and therefore each one of them involves a call to a matrix addition or subtraction procedure (just the usual method of adding, nothing fancy!). Therefore the function $MatrProd$ makes 25 calls, 7 of which are recursively to itself, and 18 of which are to a matrix addition/subtraction routine.

We will now study the complexity of the routine in two ways. We will count the number of multiplications of numbers that are needed to multiply two $2^n \times 2^n$ matrices using $MatrProd$ (call that number $f(n)$), and then we will count the number of additions of numbers (call it $g(n)$) that $MatrProd$ needs in order to multiply two $2^n \times 2^n$ matrices.

The multiplications of numbers are easy to count. $MatrProd$ calls itself 7 times, in each of which it does exactly $f(n-1)$ multiplications of numbers, hence $f(n) = 7f(n-1)$ and $f(0) = 1$ (why?). Therefore we see that $f(n) = 7^n$ for all $n \geq 0$. Hence $MatrProd$ does 7^n multiplications of numbers in order to do one multiplication of $2^n \times 2^n$ matrices.

Let's take the last sentence in the above paragraph and replace '2^n' by N throughout. It then tells us that $MatrProd$ does $7^{\log N / \log 2}$ multiplications of numbers in order to do one multiplication of $N \times N$ matrices. Since $7^{\log N / \log 2} = N^{\log 7 / \log 2} = N^{2.81\cdots}$, we see that Strassen's method uses only $O(N^{2.81})$ multiplications of numbers, in place of the N^3 such multiplications that are required by the usual formulas.

It remains to count the additions/subtractions of numbers that are needed by $MatrProd$.

In each of its 7 recursive calls to itself $MatrProd$ does $g(n-1)$ additions of numbers. In each of its 18 calls to the procedure that adds or subtracts matrices it does a number of additions of numbers that is equal to the square of the size of the matrices that are being added or subtracted. That size is 2^{n-1}, so each of the 18 such calls does 2^{2n-2} additions of numbers. It follows that $g(0) = 0$ and for $n \geq 1$ we have

$$g(n) = 7g(n-1) + 18 \cdot 4^{n-1}$$
$$= 7g(n-1) + \frac{9}{2}4^n.$$

We follow the method of section 1.4 on this first-order linear difference equation. Hence we make the change of variable $g(n) = 7^n y_n \quad (n \geq 0)$ and we find that $y_0 = 0$ and for $n \geq 1$,

$$y_n = y_{n-1} + \frac{9}{2}(4/7)^n.$$

If we sum over n we obtain

$$y_n = \frac{9}{2} \sum_{j=1}^{n} (4/7)^j$$

$$\leq \frac{9}{2} \sum_{j=0}^{\infty} (4/7)^n$$

$$= 21/2.$$

Finally, $g(n) = 7^n y_n \leq (10.5)7^n = O(7^n)$, and this is $O(N^{2.81})$ as before. This completes the proof of

Theorem 2.4.1. *In Strassen's method of fast matrix multiplication the number of multiplications of numbers, of additions of numbers and of subtractions of numbers that are needed to multiply together two $N \times N$ matrices are each $O(N^{2.81})$ (in contrast to the $\Theta(N^3)$ of the conventional method).* ■

In the years that have elapsed since Strassen's original paper many researchers have been whittling away at the exponent of N in the complexity bounds. Several new, and more elaborate algorithms have been developed, and the exponent, which was originally 3, has progressed downwards through 2.81 to values below 2.5. It is widely believed that the true minimum exponent is $2 + \epsilon$, *i.e.*, that two $N \times N$ matrices can be multiplied in time $O(N^{2+\epsilon})$, but there seems to be a good deal of work to be done before that result can be achieved.

Exercises for section 2.4

1. Suppose we could multiply together two 3×3 matrices with only 22 multiplications of numbers. How fast, recursively, would we then be able to multiply two $N \times N$ matrices?

2. (cont.) With what would the '22' in problem 1 above have to be replaced in order to achieve an improvement over Strassen's algorithm given in the text?

3. (cont.) Still more generally, with how few multiplications would we have to be able to multiply two $M \times M$ matrices in order to insure that recursively we would then be able to multiply two $N \times N$ matrices faster than the method given in this section?

4. We showed in the text that if N is a power of 2 then two $N \times N$ matrices can be multiplied in at most time $CN^{\log_2 7}$, where C is a suitable constant. Prove that if N is not a power of 2 then two $N \times N$ matrices can be multiplied in time at most $7CN^{\log_2 7}$.

2.5 The discrete Fourier transform

It is a lot easier to multiply two numbers than to multiply two polynomials.

If you should want to multiply two polynomials f and g, of degrees 77 and 94, respectively, you are in for a lot of work. To calculate just one coefficient of the product is already a lot of work. Think about the calculation of the coefficient of x^{50} in the product, for instance, and you will see that about 50 numbers must be multiplied together and added in order to calculate just that one coefficient of fg, and there are 171 other coefficients to calculate!

Instead of calculating the *coefficients* of the product fg it would be much easier just to calculate the *values* of the product at, say, 172 points. To do that we could just multiply the values of f and of g at each of those points, and after a total cost of 172 multiplications we would have the values of the product.

The values of the product polynomial at 172 distinct points determine that polynomial completely, so that sequence of values *is* the answer. It's just that we humans prefer to see polynomials given by means of their coefficients instead of by their values.

The Fourier transform, that is the subject of this section, is a method of converting from one representation of a polynomial to another. More exactly, it converts from the sequence of *coefficients* of the polynomial to the sequence of *values* of that polynomial at a certain set of points. Ease of converting between these two representations of a polynomial is vitally important for many reasons, including multiplication of polynomials, high precision integer arithmetic in computers, creation of medical images in CAT scanners and NMR scanners, etc.

Hence, in this section we will study the discrete Fourier transform of a finite sequence of numbers, methods of calculating it, and some applications.

This is a computational problem which at first glance seems very simple. What we're asked to do, basically, is to evaluate a polynomial of degree $n - 1$ at n different points. So what could be so difficult about that?

If we just calculate the n values by brute force, we certainly won't need to do more than n multiplications of numbers to find each of the n values of the polynomial that we want, so we surely don't need more than $O(n^2)$ multiplications altogether.

The interesting thing is that this particular problem is so important, and turns up in so many different applications, that it really pays to be very efficient about how the calculation is done. We will see in this section that if we use a fairly subtle method of doing this computation instead of the obvious method, then the work can be cut down from $O(n^2)$ to $O(n \log n)$. In view of the huge arrays on which this program is often run, the saving is very much worthwhile.

One can think of the Fourier transform as being a way of changing the description, or *coding* of a polynomial, so we will introduce the subject by discussing it from that point of view.

Next we will discuss the obvious way of computing the transform.

Then we will describe the 'Fast Fourier Transform', which is a rather un-obvious, but very fast, method of computing the same creature.

Finally we will discuss an important application of the subject, to the fast multiplication of polynomials.

There are many different ways that might choose to describe ('encode') a particular polynomial. Take the polynomial $f(t) = t(6 - 5t + t^2)$, for instance. This can be uniquely described in any of the following ways (and a lot more).

It is the polynomial whose
  (i) *coefficients* are 0, 6, -5, 1 or whose
  (ii) *roots* are 0, 2 and 3, and whose highest coefficient is 1 or whose
  (iii) *values* at $t = 0, 1, 2, 3$ are 0, 2, 0, 0, respectively, or whose
  (iv) *values* at the fourth-roots of unity $1, i, -1, -i$ are $2, 5 + 5i, -12$, $5 - 5i$, or etc.

We want to focus on two of these ways of representing a polynomial. The first is by its coefficient sequence; the second is by its sequence of values at the nth roots of unity, where n is 1 more than the degree of the polynomial. The process by which we pass from the coefficient sequence to the sequence of values at the roots of unity is called forming the *Fourier transform* of the coefficient sequence. To use the example above, we would say that the Fourier transform of the sequence

$$0, 6, -5, 1 \tag{2.5.1}$$

is the sequence

$$2, 5 + 5i, -12, 5 - 5i. \tag{2.5.2}$$

In general, if we are given a sequence

$$x_0, x_1, \ldots, x_{n-1} \tag{2.5.3}$$

then we think of the polynomial

$$f(t) = x_0 + x_1 t + x_2 t^2 + \cdots + x_{n-1} t^{n-1} \tag{2.5.4}$$

and we compute its values at the nth roots of unity. These roots of unity are the numbers

$$\omega_j = e^{2\pi i j/n} \qquad (j = 0, 1, \ldots, n-1). \tag{2.5.5}$$

Consequently, if we calculate the values of the polynomial (2.5.4) at the n numbers (2.5.5), we find the Fourier transform of the given sequence (2.5.3) to be the sequence

$$
\begin{aligned}
f(\omega_j) &= \sum_{k=0}^{n-1} x_k {\omega_j}^k \\
&= \sum_{k=0}^{n-1} x_k e^{2\pi i j k/n} \qquad (j = 0, 1, \ldots n-1).
\end{aligned}
\tag{2.5.6}
$$

Before proceeding, the reader should pause for a moment and make sure that the fact that (2.5.1)-(2.5.2) is a special case of (2.5.3)-(2.5.6) is clearly understood. The Fourier transform of a sequence of n numbers is another sequence of n numbers, namely the sequence of values at the nth roots of unity of the very same polynomial whose coefficients are the members of the original sequence.

The Fourier transform moves us from *coefficients* to *values at roots of unity*. Some good reasons for wanting to make that trip will appear presently, but for the moment, let's consider the computational side of the question, namely how to compute the Fourier transform efficiently.

We are going to derive an elegant and very speedy algorithm for the evaluation of Fourier transforms. The algorithm is called the Fast Fourier Transform (FFT) algorithm. In order to appreciate how fast it is, let's see how long it would take us to calculate the transform without any very clever procedure.

What we have to do is to compute the values of a given polynomial at n given points. How much work is required to calculate the value of a polynomial at *one* given point? If we want to calculate the value of the polynomial $x_0 + x_1t + x_2t^2 + \ldots + x_{n-1}t^{n-1}$ at exactly one value of t, then we can do (think how you would do it, before looking)

```
function value(x :coeff array; n:integer; t:complex);
{computes value := x₀ + x₁t + ··· + xₙ₋₁tⁿ⁻¹}
value := 0;
for j := n − 1 to 0 step −1 do
    value := t · value + xⱼ
end.{value}
```

This well-known algorithm (= 'synthetic division') for computing the value of a polynomial at a single point t obviously runs in time $O(n)$.

If we calculate the Fourier transform of a given sequence of n points by calling the function *value* n times, once for each point of evaluation, then obviously we are looking at a simple algorithm that requires $\Theta(n^2)$ time.

With the FFT we will see that the whole job can be done in time $O(n \log n)$, and we will then look at some implications of that fact. To put it another way, the cost of calculating all n of the values of a polynomial f at the nth roots of unity is much less than n times the cost of one such calculation.

First we consider the important case where n is a power of 2, say $n = 2^r$. Then the values of f, a polynomial of degree $2^r - 1$, at the $(2^r)^{th}$ roots of unity are, from (2.5.6),

$$f(\omega_j) = \sum_{k=0}^{n-1} x_k exp\{2\pi ijk/2^r\} \qquad (j = 0, 1, \ldots, 2^r - 1). \qquad (2.5.7)$$

Let's break up the sum into two sums, containing respectively the terms where k is even and those where k is odd. In the first sum write $k = 2m$ and in the second put $k = 2m + 1$. Then, for each $j = 0, 1, \ldots, 2^r - 1$,

$$
\begin{aligned}
f(\omega_j) &= \sum_{m=0}^{2^{r-1}-1} x_{2m} e^{2\pi ijm/2^{r-1}} + \sum_{m=0}^{2^{r-1}-1} x_{2m+1} e^{2\pi ij(2m+1)/2^r} \\
&= \sum_{m=0}^{2^{r-1}-1} x_{2m} e^{2\pi ijm/2^{r-1}} + e^{2\pi ij/2^r} \sum_{m=0}^{2^{r-1}-1} x_{2m+1} e^{2\pi ijm/2^{r-1}}.
\end{aligned}
$$
$$(2.5.8)$$

Something special just happened. Each of the two sums that appear in the last member of (2.5.8) is itself a Fourier transform, of a shorter sequence. The first sum is the transform of the array

$$
x[0], x[2], x[4], \ldots, x[2^r - 2] \tag{2.5.9}
$$

and the second sum is the transform of

$$
x[1], x[3], x[5], \ldots, x[2^r - 1]. \tag{2.5.10}
$$

The stage is set (well, almost set) for a recursive program.

There is one small problem, though. In (2.5.8) we want to compute $f(\omega_j)$ for 2^r values of j, namely for $j = 0, 1, \ldots, 2^r - 1$. However, the Fourier transform of the shorter sequence (2.5.9) is defined for only 2^{r-1} values of j, namely for $j = 0, 1, \ldots, 2^{r-1} - 1$. So if we calculate the first sum by a recursive call, then we will need its values for j's that are outside the range for which it was computed.

This problem is no sooner recognized than solved. Let $Q(j)$ denote the first sum in (2.5.8). Then we claim that $Q(j)$ is a *periodic* function of j, of period 2^{r-1}, because

$$
\begin{aligned}
Q(j + 2^{r-1}) &= \sum_{m=0}^{2^{r-1}-1} x_{2m} exp\{2\pi im(j + 2^{r-1})/2^{r-1}\} \\
&= \sum_{m=0}^{2^{r-1}-1} x_{2m} exp\{2\pi imj/2^{r-1}\} e^{2\pi im} \\
&= \sum_{m=0}^{2^{r-1}-1} x_{2m} exp\{2\pi imj/2^{r-1}\} \\
&= Q(j)
\end{aligned}
$$
$$(2.5.11)$$

87

for all integers j. If $Q(j)$ has been computed only for $0 \le j \le 2^{r-1} - 1$ and if we should want its value for some $j \ge 2^{r-1}$ then we can get that value by asking for $Q(j \bmod 2^{r-1})$.

Now we can state the recursive form of the Fast Fourier Transform algorithm in the (most important) case where n is a power of 2. In the algorithm we will use the type *complexarray* to denote an array of complex numbers.

> function $FFT(n$:integer; $\mathbf{x}$:complexarray):complexarray;
> {computes fast Fourier transform of $n = 2^k$ numbers $\mathbf{x}$ }
>   if $n = 1$ then $FFT[0] := x[0]$
>   **else**
>   *evenarray* $:= \{x[0], x[2], \ldots, x[n-2]\}$;
>   *oddarray* $:= \{x[1], x[3], \ldots, x[n-1]\}$;
> $\{u[0], u[1], \ldots u[\frac{n}{2}-1]\} := FFT(n/2, evenarray)$;
> $\{v[0], v[1], \ldots v[\frac{n}{2}-1]\} := FFT(n/2, oddarray)$;
>   **for** $j := 0$ **to** $n-1$ **do**
> $\tau := exp\{2\pi ij/n\}$;
> $FFT[j] := u[j \bmod \frac{n}{2}] + \tau v[j \bmod \frac{n}{2}]$
>   end.$\{FFT\}$

Let $y(k)$ denote the number of multiplications of complex numbers that will be done if we call FFT on an array whose length is $n = 2^k$. The call to $FFT(n/2, evenarray)$ costs $y(k-1)$ multiplications as does the call to $FFT(n/2, oddarray)$. The 'for $j:= 0$ to n' loop requires n more multiplications. Hence

$$y(k) = 2y(k-1) + 2^k \qquad (k \ge 1;\ y(0) = 0). \qquad (2.5.12)$$

If we change variables by writing $y(k) = 2^k z_k$, then we find that $z_k = z_{k-1} + 1$, which, together with $z_0 = 0$, implies that $z_k = k$ for all $k \ge 0$, and therefore that $y(k) = k2^k$. This proves

Theorem 2.5.1. *The Fourier transform of a sequence of n complex numbers is computed using only $O(n \log n)$ multiplications of complex numbers by means of the procedure FFT, if n is a power of 2.*

Next* we will discuss the situation when n is not a power of 2.

---

* The remainder of this section can be omitted at a first reading.

The reader may observe that by 'padding out' the input array with additional 0's we can extend the length of the array until it becomes a power of 2, and then call the *FFT* procedure that we have already discussed. In a particular application, that may or may not be acceptable. The problem is that the original question asked for the values of the input polynomial at the nth roots of unity, but after the padding, we will find the values at the Nth roots of unity, where N is the next power of 2. In some applications, such as the multiplication of polynomials that we will discuss later in this section, that change is acceptable, but in others the substitution of Nth roots for n^{th} roots may not be permitted.

We will suppose that the FFT of a sequence of n numbers is wanted, where n is not a power of 2, and where the padding operation is not acceptable. If n is a prime number we will have nothing more to say, *i.e.* we will not discuss any improvements to the obvious method for calculating the transform, one root of unity at a time.

Suppose that n is not prime (n is 'composite'). Then we can factor the integer n in some nontrivial way, say $n = r_1 r_2$ where neither r_1 nor r_2 is 1.

We claim, then, that the Fourier transform of a sequence of length n can be computed by recursively finding the Fourier transforms of r_1 different sequences, each of length r_2. The method is a straightforward generalization of the idea that we have already used in the case where n was a power of 2.

In the following we will write $\xi_n = e^{2\pi i/n}$. The train of '=' signs in the equation below shows how the question on an input array of length n is changed into r_1 questions about input arrays of length r_2. We have, for the value of the input polynomial f at the jth one of the n n^{th} roots of unity, the relations

$$
\begin{aligned}
f(e^{2\pi i j/n}) &= \sum_{s=0}^{n-1} x_s {\xi_n}^{js} \\
&= \sum_{k=0}^{r_1-1} \sum_{t=0}^{r_2-1} \{ x_{tr_1+k} {\xi_n}^{j(tr_1+k)} \} \\
&= \sum_{k=0}^{r_1-1} \sum_{t=0}^{r_2-1} \{ x_{tr_1+k} {\xi_n}^{tjr_1} {\xi_n}^{kj} \} \qquad (2.5.13) \\
&= \sum_{k=0}^{r_1-1} \{ \sum_{t=0}^{r_2-1} x_{tr_1+k} {\xi_{r_2}}^{tj} \} {\xi_n}^{kj} \\
&= \sum_{k=0}^{r_1-1} a_k(j) {\xi_n}^{kj}.
\end{aligned}
$$

We will discuss (2.5.13), line-by-line. The first '=' sign is the definition of the jth entry of the Fourier transform of the input array $\mathbf{x}$. The second equality uses the fact that every integer s such that $0 \le s \le n-1$ can be uniquely written in the form $s = tr_1 + k$, where $0 \le t \le r_2 - 1$ and $0 \le k \le r_1 - 1$. The next '=' is just a rearrangement, but the next one uses the all-important fact that ${\xi_n}^{r_1} = \xi_{r_2}$ (why?), and in the last equation we are simply defining a set of numbers

$$
a_k(j) = \sum_{t=0}^{r_2-1} x_{tr_1+k} {\xi_{r_2}}^{tj} \qquad (0 \le k \le r_1 - 1; 0 \le j \le n - 1). \qquad (2.5.14)
$$

The important thing to notice is that for a fixed k the numbers $a_k(j)$ are *periodic* in n, of period r_2, i.e., that $a_k(j + r_2) = a_k(j)$ for all j. Hence, even though the values of the $a_k(j)$ are needed for $j = 0, 1, \ldots, n - 1$, they must be computed only for $j = 0, 1, \ldots, r_2 - 1$.

Now the entire job can be done recursively, because for fixed k the set of values of $a_k(j)$ $(j = 0, 1, \ldots, r_2 - 1)$ that we must compute is itself a Fourier transform, namely of the sequence

$$
\{ x_{tr_1+k} \} \qquad (t = 0, 1, \ldots, r_2 - 1). \qquad (2.5.15)
$$

Let $g(n)$ denote the number of complex multiplications that are needed to compute the Fourier transform of a sequence of n numbers. Then, for k fixed we can recursively compute the r_2 values of $a_k(j)$ that we need with

$g(r_2)$ multiplications of complex numbers. There are r_1 such fixed values of k for which we must do the computation, hence all of the necessary values of $a_k(j)$ can be found with $r_1 g(r_2)$ complex multiplications. Once the $a_k(j)$ are all in hand, then the computation of the one value of the transform from (2.5.13) will require an additional $r_1 - 1$ complex multiplications. Since $n = r_1 r_2$ values of the transform have to be computed, we will need $r_1 r_2 (r_1 - 1)$ complex multiplications.

The complete computation needs $r_1 g(r_2) + r_1^2 r_2 - r_1 r_2$ multiplications if we choose a particular factorization $n = r_1 r_2$. The factorization that should be chosen is the one that minimizes the labor, so we have the recurrence

$$g(n) = \min_{n = r_1 r_2} \{ r_1 g(r_2) + r_1^2 r_2 \} - n. \qquad (2.5.16)$$

If $n = p$ is a prime number then there are no factorizations to choose from and our algorithm is no help at all. There is no recourse but to calculate the p values of the transform directly from the definition (2.5.6), and that will require $p - 1$ complex multiplications to be done in order to get each of those p values. Hence we have, in addition to the recurrence formula (2.5.16), the special values

$$g(p) = p(p - 1) \qquad \text{(if p is prime).} \qquad (2.5.17)$$

The recurrence formula (2.5.16) together with the starting values that are shown in (2.5.17) completely determine the function $g(n)$. Before proceeding, the reader is invited to calculate $g(12)$ and $g(18)$.

We are going to work out the exact solution of the interesting recurrence (2.5.16), (2.5.17), and when we are finished we will see which factorization of n is the best one to choose. If we leave that question in abeyance for a while, though, we can summarize by stating the (otherwise) complete algorithm for the fast Fourier transform.

> function FFT(x:complexarray; n:integer):complexarray;
> {computes Fourier transform of a sequence x of length n}
> if n is prime
> **then**
>   **for** j:=0 **to** $n-1$ **do**
> $FFT[j] := \sum_{k=0}^{n-1} x[k]\xi_n^{jk}$
> **else**
>   let $n = r_1 r_2$ be some factorization of n;
>   {see below for best choice of r_1, r_2}
>   **for** k:=0 **to** r_1-1 **do**
> $\{a_k[0], a_k[1], \ldots, a_k[r_2-1]\}$
> $:= FFT(\{x[k], x[k+r_1], \ldots, x[k+(r_2-1)r_1]\}, r_2);$
>   **for** j:=0 **to** $n-1$ **do**
> $FFT[j] := \sum_{k=0}^{r_1-1} a_k[j \bmod r_2]\xi_n^{kj}$
> end.$\{FFT\}$

Our next task will be to solve the recurrence relations (2.5.16), (2.5.17), and thereby to learn the best choice of the factorization of n.

Let $g(n) = nh(n)$, where h is a new unknown function. Then the recurrence that we have to solve takes the form

$$h(n) = \begin{cases} \min_d\{h(n/d) + d\} - 1, & \text{if } n \text{ is composite}; \\ n - 1, & \text{if } n \text{ is prime}. \end{cases} \qquad (2.5.18)$$

In (2.5.18), the 'min' is taken over all d that divide n other than $d = 1$ and $d = n$.

The above relation determines the value of h for all positive integers n. For example,

$$h(15) = \min_d(h(15/d) + d) - 1$$
$$= \min(h(5) + 3, h(3) + 5) - 1$$
$$= \min(7, 7) - 1 = 6$$

and so forth.

To find the solution in a pleasant form, let

$$n = p_1^{a_1} p_2^{a_2} \cdots p_s^{a_s} \qquad (2.5.19)$$

be the canonical factorization of n into primes. We claim that the function

$$h(n) = a_1(p_1 - 1) + a_2(p_2 - 1) + \cdots + a_s(p_s - 1) \qquad (2.5.20)$$

is the solution of (2.5.18) (this claim is obviously (?) correct if n is prime).

To prove the claim in general, suppose it to be true for $1, 2, \ldots, n-1$, and suppose that n is not prime. Then every divisor d of n must be of the form $d = p_1^{b_1} p_2^{b_2} \cdots p_s^{b_s}$, where the primes p_i are the same as those that appear in (2.5.19) and each b_i is $\leq a_i$. Hence from (2.5.18) we get

$$h(n) = \min_{\mathbf{b}}\{(a_1 - b_1)(p_1 - 1) + \cdots + (a_s - b_s)(p_s - 1) + p_1^{b_1} \cdots p_s^{b_s}\} - 1 \quad (2.5.21)$$

where now the 'min' extends over all admissible choices of the b's, namely exponents $b_1, \ldots, b_s$ such that $0 \leq b_i \leq a_i$ $(\forall i = 1, s)$ and not all b_i are 0 and not all $b_i = a_i$.

One such admissible choice would be to take, say, $b_j = 1$ and all other $b_i = 0$. If we let $H(b_1, \ldots, b_s)$ denote the quantity in braces in (2.5.21), then with this choice the value of H would be $a_1(p_1 - 1) + \cdots + a_s(p_s - 1) + 1$, exactly what we need to prove our claim (2.5.20). Hence what we have to show is that the above choice of the b_i's is the best one. We will show that if one of the b_i is larger than 1 then we can reduce it without increasing the value of H.

To prove this, observe that for each $i = 1, s$ we have

$$H(b_1, \ldots, b_i + 1, \ldots, b_s) - H(b_1, \ldots, b_s) = -p_i + d(p_i - 1)$$
$$= (d - 1)(p_i - 1).$$

Since the divisor $d \geq 2$ and the prime $p_i \geq 2$, the last difference is nonnegative. Hence H doesn't increase if we decrease one of the b's by 1 unit, as long as not all $b_i = 0$. It follows that the minimum of H occurs among the *prime* divisors d of n. Further, if d is prime, then we can easily check from (2.5.21) that it doesn't matter which prime divisor of n that we choose to be d, the function $h(n)$ is always given by (2.5.20). If we recall the change of variable $g(n) = nh(n)$ we find that we have proved

Theorem 2.5.2. *(Complexity of the Fast Fourier Transform) The best choice of the factorization $n = r_1 r_2$ in algorithm FFT is to take r_1 to be a prime divisor of n. If that is done, then algorithm FFT requires*

$$g(n) = n(a_1(p_1 - 1) + a_2(p_2 - 1) + \cdots + a_s(p_s - 1))$$

complex multiplications in order to do its job, where $n = p_1^{a_1} \cdots p_s^{a_s}$ is the canonical factorization of the integer n. ∎

Table 2.5.1 shows the number $g(n)$ of complex multiplications required by *FFT* as a function of n. The saving over the straightforward algorithm that uses $n(n-1)$ multiplications for each n is apparent.

n	g(n)	n	g(n)
2	2	22	242
3	6	23	506
4	8	24	120
5	20	25	200
6	18	26	338
7	42	27	162
8	24	28	224
9	36	29	812
10	50	30	210
11	110	31	930
12	48	32	160
13	156	33	396
14	98	34	578
15	90	35	350
16	64	36	216
17	272	37	1332
18	90	38	722
19	342	39	546
20	120	40	280
21	168	41	1640

Table 2.5.1: The complexity of the FFT

If n is a power of 2, say $n = 2^q$, then the formula of theorem 2.5.2 reduces to $g(n) = n \log n / \log 2$, in agreement with theorem 2.5.1. What does the formula say if n is a power of 3? if n is a product of distinct primes?

2.6 Applications of the FFT

Finally, we will discuss some applications of the FFT. A family of such applications begins with the observation that the FFT provides the fastest game in town for multiplying two polynomials together. Consider a multiplication like

$$(1 + 2x + 7x^2 - 2x^3 - x^4) \cdot (4 - 5x - x^2 - x^3 + 11x^4 + x^5).$$

We will study the amount of labor that is needed to do this multiplication by the straightforward algorithm, and then we will see how the FFT can help.

If we do this multiplication in the obvious way then there is quite a bit of work to do. The coefficient of x^4 in the product, for instance, is $1 \cdot 11 + 2 \cdot (-1) + 7 \cdot (-1) + (-2) \cdot (-5) + (-1) \cdot 4 = 8$, and 5 multiplications are needed to compute just that single coefficient of the product polynomial.

In the general case, we want to multiply

$$\{\sum_{i=0}^{n} a_i x^i\} \cdot \{\sum_{j=0}^{m} b_j x^j\}. \tag{2.6.1}$$

In the product polynomial, the coefficient of x^k is

$$\sum_{r=\max(0,k-m)}^{\min(k,n)} a_r b_{k-r}. \tag{2.6.2}$$

For k fixed, the number of terms in the sum (2.6.2) is $\min(k,n) - \max(0, k - m) + 1$. If we sum this amount of labor over $k = 0, m + n$ we find that the total amount of labor for multiplication of two polynomials of degrees m and n is $\Theta(mn)$. In particular, if the polynomials are of the same degree n then the labor is $\Theta(n^2)$.

By using the FFT the amount of labor can be reduced from $\Theta(n^2)$ to $\Theta(n \log n)$.

To understand how this works, let's recall the definition of the Fourier transform of a sequence. It is the sequence of values of the polynomial whose coefficients are the given numbers, at the nth roots of unity, where n is the length of the input sequence.

Imagine two universes, one in which the residents are used to describing polynomials by means of their coefficients, and another one in which the inhabitants are fond of describing polynomials by their values at roots of unity. In the first universe the locals have to work fairly hard to multiply two polynomials because they have to carry out the operations (2.6.2) in order to find each coefficient of the product.

In the second universe, multiplying two polynomials is a breeze. If we have in front of us the values $f(\omega)$ of the polynomial f at the roots of unity, and the values $g(\omega)$ of the polynomial g at the same roots of unity, then what are the values $(fg)(\omega)$ of the product polynomial fg at the roots of unity? To find each one requires only a single multiplication of two complex numbers, because the value of fg at ω is simply $f(\omega)g(\omega)$.

Multiplying values is easier than finding the coefficients of the product.

Since we live in a universe where people like to think about polynomials as being given by their coefficient arrays, we have to take a somewhat roundabout route in order to do an efficient multiplication.

Given: A polynomial f, of degree n, and a polynomial g of degree m; by their coefficient arrays. Wanted: The coefficients of the product polynomial fg, of degree $m + n$.

Step 1. Let $N - 1$ be the smallest integer that is a power of 2 and is greater than $m + n + 1$.

Step 2. Think of f and g as polynomials each of whose degrees is $N - 1$. This means that we should adjoin $N - n$ more coefficients, all $= 0$, to the coefficient array of f and $N - m$ more coefficients, all $= 0$, to the coefficient array of g. Now both input coefficient arrays are of length N.

Step 3. Compute the FFT of the array of coefficients of f. Now we are looking at the values of f at the Nth roots of unity. Likewise compute the FFT of the array of coefficients of g to obtain the array of values of g at the same N^{th} roots of unity. The cost of this step is $O(N \log N)$.

Step 4. For each of the Nth roots of unity ω multiply the number $f(\omega)$ by the number $g(\omega)$. We now have the numbers $f(\omega)g(\omega)$, which are exactly the values of the unknown product polynomial fg at the Nth roots of unity. The cost of this step is N multiplications of numbers, one for each ω.

Step 5. We now are looking at the *values* of fg at the Nth roots, and we want to get back to the *coefficients* of fg because that was what we were asked for. To go backwards, from values at roots of unity to coefficients, calls for the *inverse Fourier transform*, which we will describe in a moment.

Its cost is also $O(N \log N)$. ∎

The answer to the original question has been obtained at a total cost of $O(N \log N) = O((m + n) \log (m + n))$ arithmetic operations. It's true that we did have to take a walk from our universe to the next one and back again, but the round trip was a lot cheaper than the $O((m + n)^2)$ cost of a direct multiplication.

It remains to discuss the inverse Fourier transform. Perhaps the neatest way to do that is to juxtapose the formulas for the Fourier transform and for the inverse tranform, so as to facilitate comparison of the two, so here they are. If we are given a sequence $\{x_0, x_1, \ldots, x_{n-1}\}$ then the Fourier transform of the sequence is the sequence (see (2.5.6))

$$f(\omega_j) = \sum_{k=0}^{n-1} x_k e^{2\pi i j k / n} \qquad (j = 0, 1, \ldots, n - 1). \qquad (2.6.3)$$

Conversely, if we are given the numbers $f(\omega_j)$ $(j = 0, \ldots, n - 1)$ then we can recover the coefficient sequence $x_0, \ldots, x_{n-1}$ by the inverse formulas

$$x_k = \frac{1}{n} \sum_{j=0}^{n-1} f(\omega_j) e^{-2\pi i j k / n} \qquad (k = 0, 1, \ldots, n - 1). \qquad (2.6.4)$$

The differences between the inverse formulas and the original transform formulas are first the appearance of '$1/n$' in front of the summation and second the '$-$' sign in the exponential. We leave it as an exercise for the reader to verify that these formulas really do invert each other.

We observe that if we are already in possession of a computer program that will find the FFT, then we can use it to calculate the inverse Fourier transform as follows:

(i) Given a sequence $\{f(\omega)\}$ of values of a polynomial at the nth roots of unity, form the complex conjugate of each member of the sequence.

(ii) Input the conjugated sequence to your FFT program.

(iii) Form the complex conjugate of each entry of the output array, and divide by n. You now have the inverse transform of the input sequence.

The cost is obviously equal to the cost of the FFT plus a linear number of conjugations and divisions by n.

An outgrowth of the rapidity with which we can now multiply polynomials is a rethinking of the methods by which we do ultrahigh-precision arithmetic. How fast can we multiply two integers, each of which has ten million bits? By using ideas that developed directly (though not at all trivially) from the ones that we have been discussing, Schönhage and Strassen found the fastest known method for doing such large-scale multiplications of integers. The method relies heavily on the FFT, which may not be too surprising since an integer n is given in terms of its bits $b_0, b_1, \ldots, b_m$ by the relation

$$n = \sum_{i \geq 0} b_i 2^i. \tag{2.6.5}$$

However the sum in (2.6.5) is seen at once to be the value of a certain polynomial at $x = 2$. Hence in asking for the bits of the product of two such integers we are asking for something very similar to the coefficients of the product of two polynomials, and indeed the fastest known algorithms for this problem depend upon the Fast Fourier Transform.

Exercises for section 2.6

1. Let ω be an nth root of unity, and let k be a fixed integer. Evaluate

$$1 + \omega^k + \omega^{2k} + \cdots + \omega^{k(n-1)}.$$

2. Verify that the relations (2.6.3) and (2.6.4) indeed are inverses of each other.

3. Let $f = \sum_{j=0}^{n-1} a_j x^j$. Show that

$$\frac{1}{n} \sum_{\omega^n = 1} |f(\omega)|^2 = |a_0|^2 + \cdots + |a_{n-1}|^2$$

4. The values of a certain cubic polynomial at $1, i, -1, -i$ are $1, 2, 3, 4$, respectively. Find its value at 2.

5. Write a program that will do the FFT in the case where the number of data points is a power of 2. Organize your program so as to minimize additional array storage beyond the input and output arrays.

6. Prove that a polynomial of degree n is uniquely determined by its values at $n + 1$ distinct points.

2.7 A review

Here is a quick review of the algorithms that we studied in this chapter.

Sorting is an easy computational problem. The most obvious way to sort n array elements takes time $\Theta(n^2)$. We discussed a recursive algorithm that sorts in an average time of $\Theta(n \log n)$.

Finding a maximum independent set in a graph is a hard computational problem. The most obvious way to find one might take time $\Theta(2^n)$ if the graph G has n vertices. We discussed a recursive method that runs in time $\Theta(1.39^n)$. The best known methods run in time $\Theta(2^{n/4})$.

Finding out if a graph is K-colorable is a hard computational problem. The most obvious way to do it takes time $\Theta(K^n)$, if G has n vertices. We discussed a recursive method that runs in time $O(1.62^{n+E})$ if G has n vertices and E edges. One recently developed method * runs in time $O((1 + \sqrt[3]{3})^n)$. We will see in section 5.7 that this problem can be done in an *average* time that is $O(1)$ for fixed K.

Multiplying two matrices is an easy computational problem. The most obvious way to do it takes time $\Theta(n^3)$ if the matrices are $n \times n$. We discussed a recursive method that runs in time $O(n^{2.82})$. A recent method ** runs in time $O(n^\gamma)$ for some $\gamma < 2.5$.

Finding the discrete Fourier transform of an array of n elements is an easy computational problem. The most obvious way to do it takes time $\Theta(n^2)$. We discussed a recursive method that runs in time $O(n \log n)$ if n is a power of 2.

When we write a program recursively we are making life easier for ourselves and harder for the compiler and the computer. A single call to a recursive program can cause it to execute a tree-full of calls to itself before it is able to respond to our original request.

For example, if we call Quicksort to sort the array

$$\{5, 8, 13, 9, 15, 29, 44, 71, 67\}$$

then the tree shown in Fig. 2.7.1 might be generated by the compiler.

---

* E. Lawler, A note on the complexity of the chromatic number problem, *Information Processing Letters* **5** (1976), 66-7.

** D. Coppersmith and S. Winograd, On the asymptotic complexity of matrix multiplication, *SIAM J. Comp.* **11** (1980), 472-492.

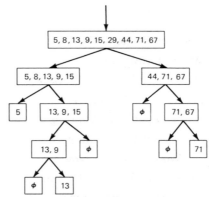

Fig. 2.7.1: A tree of calls to *Quicksort*

Again, if we call *maxset1* on the 5-cycle, the tree in Fig. 2.3.3 of calls may be created.

A single invocation of *chrompoly*, where the input graph is a 4-cycle, for instance, might generate the tree of recursive calls that appears in Fig. 2.7.2.

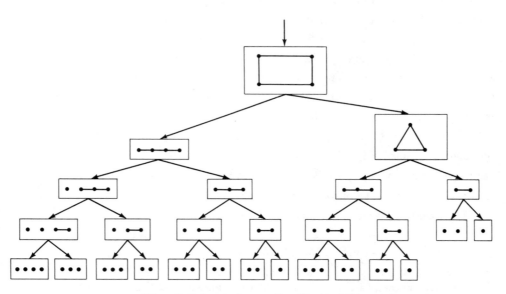

Fig. 2.7.2: A tree of calls to *chrompoly*

Finally, if we call the 'power of 2' version of the *FFT* algorithm on the sequence $\{1, i, -i, 1\}$ then *FFT* will proceed to manufacture the tree shown in Fig. 2.7.3.

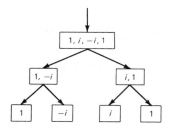

Fig. 2.7.3: The recursive call tree for *FFT*

It must be emphasized that the creation of the tree of recursions is done by the compiler without any further effort on the part of the programmer. As long as we're here, how does a compiler go about making such a tree?

It does it by using an auxiliary *stack*. It adopts the philosophy that if it is asked to do two things at once, well after all, it can't do that, so it does one of those two things and drops the other request on top of a stack of unfinished business. When it finishes executing the first request it goes to the top of the stack to find out what to do next.

Example

Let's follow the compiler through its tribulations as it attempts to deal with our request for maximum independent set size that appears in Fig. 2.3.3. We begin by asking for the *maxset*1 of the 5-cycle. Our program immediately makes two recursive calls to *maxset*1, on each of the two graphs that appear on the second level of the tree in Fig. 2.3.3. The stack is initially empty.

The compiler says to itself 'I can't do these both at once', and it puts the right-hand graph (involving vertices 3,4) on the stack, and proceeds to call itself on the left hand graph (vertices 2,3,4,5).

When it tries to do that one, of course, two more graphs are generated, of which the right-hand one (4,5) is dropped onto the stack, on top of the graph that previously lived there, so now two graphs are on the stack, awaiting processing, and the compiler is dealing with the graph (3,4,5).

This time the graph of just one vertex (5) is dropped onto the stack, which now holds three graphs, as the compiler works on (4,5).

Next, that graph is broken up into (5), and an empty graph, which is dutifully dropped onto the stack, so the compiler can work on (5).

Finally, something fruitful happens: the graph (5) has no edges, so the program *maxset*1 gives, in its trivial case, very specific instructions as to how to deal with this graph. We now know that the graph that consists of just the single vertex (5) has a *maxset*1 values of 1.

The compiler next reaches for the graph on top of the stack, finds that it is the empty graph, which has no edges at all, and therefore its *maxset* size is 0.

It now knows the $n_1 = 1$ and the $n_2 = 0$ values that appear in the algorithm *maxset*1, and therefore it can execute the instruction *maxset*1 := $max(n_1, 1 + n_2)$, from which it finds that the value of *maxset*1 for the graph (4,5) is 1, and it continues from there, to dig itself out of the stack of unfinished business.

In general, if it is trying to execute *maxset*1 on a graph that has edges, it will drop the graph $G - \{v^*\} - Nbhd(v^*)$ on the stack and try to do the graph $G - \{v^*\}$.

The reader should try to write out, as a formal algorithm, the procedure that we have been describing, whereby the compiler deals with a recursive computation that branches into two sub-computations until a trivial case is reached. ∎

Exercise for section 2.7

1. In Fig. 2.7.3, add to the picture the *output* that each of the recursive calls gives back to the box above it that made the call.

Bibliography

A definitive account of all aspects of sorting is in

D. E. Knuth, *The art of computer programming*, Vol. 3: *Sorting and searching*, Addison Wesley, Reading MA, 1973.

All three volumes of the above reference are highly recommended for the study of algorithms and discrete mathematics.

A $O(2^{n/3})$ algorithm for the maximum independent set problem can be found in

R. E. Tarjan and A. Trojanowski, Finding a maximum independent set, *SIAM J. Computing* 6 (1977), 537-546.

Recent developments in fast matrix multiplication are traced in

Victor Pan, How to multiply matrices faster, Lecture notes in computer science No. 179, Springer-Verlag, 1984.

The realization that the Fourier transform calculation can be speeded up has been traced back to

C. Runge, *Zeits. Math. Phys.*, **48** (1903) p. 443.

and also appears in

C. Runge and H. König, *Die Grundlehren der math. Wissensch.*, 11, Springer Verlag, Berlin 1924.

The introduction of the method in modern algorithmic terms is generally credited to

J. M. Cooley and J. W. Tukey, An algorithm for the machine calculation of complex Fourier series, *Mathematics of Computation*, **19** (1965), 297-301.

A number of statistical applications of the method are in

J. M. Cooley, P. A. W. Lewis and P. D. Welch, The Fast Fourier Transform and its application to time series analysis, in *Statistical Methods for Digital Computers*, Enslein, Ralston and Wilf eds., John Wiley & Sons, New York, 1977, 377-423.

The use of the FFT for high precision integer arithmetic is due to

A Schönhage and V. Strassen, Schnelle Multiplikation grosser Zahlen, *Computing*, **7** (1971), 281-292.

An excellent account of the above as well as of applications of the FFT to polynomial arithmetic is by

A. V. Aho, J. E. Hopcroft and J. D. Ullman, The design and analysis of computer algorithms, Addison Wesley, Reading, MA, 1974 (chap. 7).

Chapter 3: The Network Flow Problem

3.1 Introduction

The network flow problem is an example of a beautiful theoretical subject that has many important applications. It also has generated algorithmic questions that have been in a state of extremely rapid development in the past 20 years. Altogether, the fastest algorithms that are now known for the problem are much faster, and some are much simpler, than the ones that were in use a short time ago, but it is still unclear how close to the 'ultimate' algorithm we are.

Definition. *A network is an edge-capacitated directed graph, with two distinguished vertices called the source and the sink.*

To repeat that, this time a little more slowly, suppose first that we are given a directed graph (*digraph*) G. That is, we are given a set of vertices, and a set of *ordered* pairs of these vertices, these pairs being the *edges* of the digraph. It is perfectly OK to have both an edge from u to v and an edge from v to u, or both, or neither, for all $u \neq v$. No edge (u, u) is permitted. If an edge e is directed *from* vertex v *to* vertex w, then v is the *initial* vertex of e and w is the *terminal* vertex of e. We may then write $v = Init(e)$ and $w = Term(e)$.

Next, in a network there is associated with each directed edge e of the digraph a positive real number called its *capacity*, and denoted by $cap(e)$.

Finally, two of the vertices of the digraph are distinguished. One, s, is the source, and the other, t, is the sink of the network.

We will let **X** denote the resulting network. It consists of the digraph G, the given set of edge capacities, the source, and the sink. A network is shown in Fig. 3.1.1.

Now roughly speaking, we can think of the edges of G as conduits for a fluid, the capacity of each edge being the carrying-capacity of the edge for that fluid. Imagine that the fluid flows in the network from the source to the sink, in such a way that the amount of fluid in each edge does not exceed the capacity of that edge.

We want to know the maximum net quantity of fluid that could be flowing from source to sink.

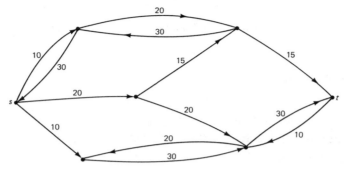

Fig. 3.1.1: A network

That was a rough description of the problem; here it is more precisely.

Definition. *A flow in a network* **X** *is a function f that assigns to each edge e of the network a real number $f(e)$, in such a way that*

(1) For each edge e we have $0 \le f(e) \le cap(e)$ and

(2) For each vertex v other than the source and the sink, it is true that

$$\sum_{Init(e)=v} f(e) = \sum_{Term(e)=v} f(e). \tag{3.1.1}$$

The condition (3.1.1) is a flow conservation condition. It states that the outflow from v (the left side of (3.1.1)) is equal to the inflow to v (the right side) for all vertices v other than s and t. In the theory of electrical networks such conservation conditions are known as Kirchhoff's laws. Flow cannot be manufactured anywhere in the network except at s or t. At other vertices, only redistribution or rerouting takes place.

Since the source and the sink are exempt from the conservation conditions there may, and usually will, be a nonzero net flow out of the source, and a nonzero net flow into the sink. Intuitively it must already be clear that these two are equal, and we will prove it below, in section 3.4. If we let Q be the net outflow from the source, then Q is also the net inflow to the sink.

The quantity Q is called the *value of the flow.*

In Fig. 3.1.2 there is shown a flow in the network of Fig. 3.1.1. The amounts of flow in each edge are shown in the square boxes. The other number on each edge is its capacity. The letter inside the small circle next to each vertex is the name of that vertex, for the purposes of the present discussion. The value of the flow in Fig. 3.1.2 is $Q = 32$.

106

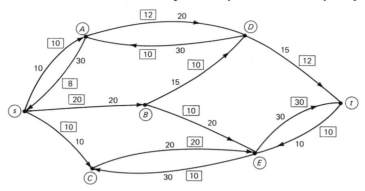

Fig. 3.1.2: A flow in a network

The *network flow problem*, the main subject of this chapter, is: *given a network* **X**, *find the maximum possible value of a flow in* **X**, *and find a flow of that value.*

3.2 Algorithms for the network flow problem

The first algorithm for the network flow problem was given by Ford and Fulkerson. They used that algorithm not only to solve instances of the problem, but also to prove theorems about network flow, a particularly happy combination. In particular, they used their algorithm to prove the 'max-flow-min-cut' theorem, which we state below as theorem 3.4.1, and which occupies a central position in the theory.

The speed of their algorithm, it turns out, depends on the edge capacities in the network as well as on the numbers V of vertices, and E of edges, of the network. Indeed, for certain (irrational) values of edge capacities they found that their algorithm might not converge at all (see section 3.5).

In 1969 Edmonds and Karp gave the first algorithm for the problem whose speed is bounded by a polynomial function of E and V only. In fact that algorithm runs in time $O(E^2 V)$. Since then there has been a steady procession of improvements in the algorithms, culminating, at the time of this writing anyway, with an $O(EV \log V)$ algorithm. The chronology is shown in Table 3.2.1.

The maximum number of edges that a network of V vertices can have is $\Theta(V^2)$. A family of networks might be called *dense* if there is a $K > 0$ such that $|E(\mathbf{X})| > K|V(\mathbf{X})|^2$ for all networks in the family. The reader should check that for dense networks, all of the time complexities in Table

Author(s)	Year	Complexity
Ford, Fulkerson	1956	$- - - -$
Edmonds, Karp	1969	$O(E^2 V)$
Dinic	1970	$O(EV^2)$
Karzanov	1973	$O(V^3)$
Cherkassky	1976	$O(\sqrt{E}V^2)$
Malhotra, *et al.*	1978	$O(V^3)$
Galil	1978	$O(V^{5/3}E^{2/3})$
Galil and Naamad	1979	$O(EV \log^2 V)$
Sleator and Tarjan	1980	$O(EV \log V)$
Goldberg and Tarjan	1985	$O(EV \log (V^2/E))$

Table 3.2.1: Progress in network flow algorithms

3.2.1, beginning with Karzanov's algorithm, are in the neighborhood of $O(V^3)$. On the other hand, for *sparse* networks (networks with relatively few edges), the later algorithms in the table will give significantly better performances than the earlier ones.

Exercise 3.2.1. *Given $K > 0$. Consider the family of all possible networks* **X** *for which* $|E(\mathbf{X})| = K|V(\mathbf{X})|$. *In this family, evaluate all of the complexity bounds in Table 3.2.1 and find the fastest algorithm for the family.*

Among the algorithms in Table 3.2.1 we will discuss just two in detail. The first will be the original algorithm of Ford and Fulkerson, because of its importance and its simplicity, if not for its speed.

The second will be the 1978 algorithm of Malhotra, Pramodh-Kumar and Maheshwari (MPM), for three reasons. It uses the idea, introduced by Dinic in 1970 and common to all later algorithms, of *layered networks*, it is fast, and it is extremely simple and elegant in its conception, and so it represents a good choice for those who may wish to program one of these algorithms for themselves.

3.3 The algorithm of Ford and Fulkerson

The basic idea of the Ford-Fulkerson algorithm for the network flow problem is this: start with some flow function (initially this might consist of zero flow on every edge). Then look for a *flow augmenting path* in the network. A flow augmenting path is a path from the source to the sink along which we can push some additional flow.

In Fig. 3.3.1 below we show a flow augmenting path for the network of Fig. 3.2.1. The capacities of the edges are shown on each edge, and the values of the flow function are shown in the boxes on the edges.

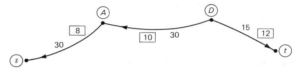

Fig. 3.3.1: A flow augmenting path

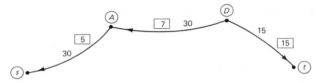

Fig. 3.3.2: The path above, after augmentation

An edge can get elected to a flow augmenting path for two possible reasons. Either

(a) the direction of the edge is *coherent* with the direction of the path from source to sink and the present value of the flow function on the edge is below the capacity of that edge, or

(b) the direction of the edge is *opposed* to that of the path from source to sink and the present value of the flow function on the edge is strictly positive.

Indeed, on all edges of a flow augmenting path that are coherently oriented with the path we can increase the flow along the edge, and on all edges that are incoherently oriented with the path we can decrease the flow on the edge, and in either case we will have *increased the value of the flow* (think about that one until it makes sense).

It is, of course, necessary to maintain the conservation of flow, *i.e.*, to respect Kirchhoff's laws. To do this we will augment the flow on every edge of an augmenting path by the same amount. If the conservation conditions were satisfied before the augmentation then they will still be satisfied after such an augmentation.

It may be helpful to remark that an edge is coherently or incoherently oriented only *with respect to a given path* from source to sink. That is, the coherence, or lack of it, is not only a property of the directed edge, but depends on how the edge sits inside a chosen path.

Thus, in Fig. 3.3.1 the first edge is directed towards the source, *i.e.*, incoherently with the path. Hence if we can *decrease* the flow in that edge we will have *increased* the value of the flow function, namely the net flow out of the source. That particular edge can indeed have its flow decreased, by at most 8 units. The next edge carries 10 units of flow towards the source. Therefore if we *decrease* the flow on that edge, by up to 10 units, we will also have *increased* the value of the flow function. Finally, the edge into the sink carries 12 units of flow and is oriented towards the sink. Hence if we *increase* the flow in this edge, by at most 3 units since its capacity is 15, we will have increased the value of the flow in the network.

Since every edge in the path that is shown in Fig. 3.3.1 can have its flow altered in one way or the other so as to increase the flow in the network, the path is indeed a flow augmenting path. The most that we might accomplish with this path would be to push 3 more units of flow through it from source to sink. We couldn't push more than 3 units through because one of the edges (the edge into the sink) will tolerate an augmentation of only 3 flow units before reaching its capacity.

To augment the flow by 3 units we would diminish the flow by 3 units on each of the first two edges and increase it by 3 units on the last edge. The resulting flow in this path is shown in Fig. 3.3.2. The flow in the full network, after this augmentation, is shown in Fig. 3.3.3. Note carefully that if these augmentations are made then flow conservation at each vertex of the network will still hold (check this!).

After augmenting the flow by 3 units as we have just described, the resulting flow will be the one that is shown in Fig. 3.3.3. The value of the flow in Fig. 3.1.2 was 32 units. After the augmentation, the flow function in Fig. 3.3.3 has a value of 35 units.

We have just described the main idea of the Ford-Fulkerson algorithm. It first finds a flow augmenting path. Then it augments the flow along that path as much as it can. Then it finds another flow augmenting path, etc. etc. The algorithm terminates when no flow augmenting paths exist. We will prove that when that happens, the flow will then be at the maximum

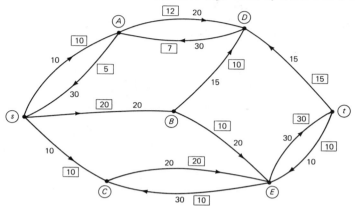

Fig. 3.3.3: The network, after augmentation of flow

possible value, *i.e.*, we will have found the solution of the network flow problem.

We will now describe the steps of the algorithm in more detail.

Definition. *Let f be a flow function in a network* **X.** *We say that an edge e of* **X** *is usable from v to w if either e is directed from v to w and the flow in e is less than the capacity of the edge, or e is directed from w to v and the flow in e is* > 0.

Now, given a network and a flow in that network, how do we find a flow augmenting path from the source to the sink? This is done by a process of labelling and scanning the vertices of the network, beginning with the source and proceeding out to the sink. Initially all vertices are in the conditions 'unlabeled' and 'unscanned.' As the algorithm proceeds, various vertices will become labeled, and if a vertex is labeled, it may become *scanned*. To scan a vertex v means, roughly, that we stand at v and look around at all neighbors w of v that haven't yet been labeled. If e is some edge that joins v with a neighbor w, and if the edge e is usable from v to w as defined above, then we will label w, because any flow augmenting path that has already reached from the source to v can be extended another step, to w.

The label that every vertex v gets is a triple $(u, \pm, z)$, and here is what the three items mean.

The 'u' part of the label of v is the name of the vertex that was being scanned when v was labeled.

The '$\pm$' will be '+' if v was labeled because the edge (u, v) was usable from u to v (*i.e.*, if the flow from u to v was less than the capacity of (u, v))

111

and it will be '−' if v was labeled because the edge (v, u) was usable from u to v (*i.e.*, if the flow from v to u was > 0).

Finally, the 'z' component of the label represents the largest amount of flow that can be pushed from the source to the present vertex v along any augmenting path that has so far been found. At each step the algorithm will replace the current value of z by the amount of new flow that could be pushed through to z along the edge that is now being examined, if that amount is smaller than z.

So much for the meanings of the various labels. As the algorithm proceeds, the labels that get attached to the different vertices form a record of how much flow can be pushed through the network from the source to the various vertices, and by exactly which routes.

To begin with, the algorithm labels the source with $(-\infty, +, \infty)$. The source now has the label-status *labeled* and the scan-status *unscanned*.

Next we will scan the source. Here is the procedure for scanning any vertex u.

> procedure *scan*(u:vertex;$\mathbf{X}$:network; f:flow);
> **for** every 'unlabeled' vertex v that is connected
> to u by an edge in either or both directions, do
> if the flow in (u, v) is less than $cap(u, v)$
> **then**
> label v with $(u, +, \min\{z(u), cap(u, v) - flow(u, v)\})$
> **else if** the flow in (v, u) is > 0
> **then**
> label v with $(u, -, \min\{z(u), flow(v, u)\})$ and
> change the label-status of v to 'labeled';
> change the scan-status of u to 'scanned'
> end.{*scan*}

We can use the above procedure to describe the complete scanning and labelling of the vertices of the network, as follows.

procedure *labelandscan*(**X** :network; f:flow; *whyhalt*:reason);
give every vertex the scan-status '*unscanned*'
    and the label-status '*unlabeled*';
$u := source$;
label *source* with $(-\infty, +, \infty)$;
label-status of *source*:= '*labeled*';
while {there is a '*labeled*' and '*unscanned*' vertex v
    and *sink* is '*unlabeled*'}
      do $scan(v, \mathbf{X}, f)$;
  if *sink* is *unlabeled*
    then '*whyhalt*':='*flow is maximum*'
    else '*whyhalt*':= '*it's time to augment*'
end.{*labelandscan*}

Obviously the labelling and scanning process will halt for one of two reasons: either the sink t acquires a label, or the sink never gets labeled but no more labels can be given. In the first case we will see that a flow augmenting path from source to sink has been found, and in the second case we will prove that the flow is at its maximum possible value, so the network flow problem has been solved.

Suppose the sink does get a label, for instance the label $(u, \pm, z)$. Then we claim that the value of the flow in the network can be augmented by z units.

To prove this we will construct a flow augmenting path, using the labels on the vertices, and then we will change the flow by z units on every edge of that path in such a way as to increase the value of the flow function by z units. This is done as follows.

If the sign part of the label of t is '+,' then increase the flow function by z units on the edge (u, t), else decrease the flow on edge (t, u) by z units.

Then move back one step away from the sink, to vertex u, and look at its label, which might be $(w, \pm, z_1)$. If the sign is '+' then increase the flow on edge (w, u) by z units (not by z_1 units!), while if the sign is '−' then decrease the flow on edge (u, w) by z units. Next replace u by w, etc., until the source s has been reached.

A little more formally, the flow augmentation algorithm is the following.

```
procedure augmentflow(X :network; f:flow ; amount:real);
{assumes that labelandscan has just been done}
v:=sink;
amount:= the 'z' part of the label of sink;
repeat
    (previous, sign, z) := label(v);
    if sign='+'
            then
            increase f(previous, v) by amount
            else
            decrease f(v, previous) by amount;
    v := previous
until v= source
end.{augmentflow}
```

The value of the flow in the network has now been *increased* by z units. The whole process of labelling and scanning is now repeated, to search for another flow augmenting path. The algorithm halts only when we are unable to label the sink. The complete Ford-Fulkerson algorithm is shown below.

```
procedure fordfulkerson(X :network; f: flow; maxflowvalue:real);
{finds maximum flow in a given network X }
  set f:=0 on every edge of X ;
  maxflowvalue:=0;
  repeat
    labelandscan(X, f, whyhalt);
    if whyhalt='it's time to augment' then
        augmentflow(X,f, amount);
        maxflowvalue := maxflowvalue + amount
  until whyhalt = 'flow is maximum'
end.{fordfulkerson}
```

Let's look at what happens if we apply the labelling and scanning algorithm to the network and flow shown in Fig. 3.1.2. First vertex s gets

114

the label $(-\infty, +, \infty)$. We then scan s. Vertex A gets the label $(s, -, 8)$, B cannot be labeled, and C gets labeled with $(s, +, 10)$, which completes the scan of s.

Next we scan vertex A, during which D acquires the label $(A, +, 8)$. Then C is scanned, which results in E getting the label $(C, -, 10)$. Finally, the scan of D results in the label $(D, +, 3)$ for the sink t.

From the label of t we see that there is a flow augmenting path in the network along which we can push 3 more units of flow from s to t. We find the path as in procedure *augmentflow* above, following the labels backwards from t to D, A and s. The path in question will be seen to be exactly the one shown in Fig. 3.3.1, and further augmentation proceeds as we have discussed above.

3.4 The max-flow min-cut theorem

Now we are going to look at the state of affairs that holds when the flow augmentation procedure terminates because it has not been able to label the sink. We want to show that then the flow will have a maximum possible value.

Let $W \subset V(\mathbf{X})$, and suppose that W contains the source and W does not contain the sink. Let $\overline{W}$ denote all other vertices of $\mathbf{X}$, *i.e.*, $\overline{W} = V(\mathbf{X}) - W$.

Definition. *By the cut $(W, \overline{W})$ we mean the set of all edges of $\mathbf{X}$ whose initial vertex is in W and whose terminal vertex is in $\overline{W}$.*

For example, one cut in a network consists of all edges whose initial vertex is the source.

Now, every unit of flow that leaves the source and arrives at the sink must at some moment flow from a vertex of W to a vertex of $\overline{W}$, *i.e.*, must flow along some edge of the cut $(W, \overline{W})$. If we define *the capacity of a cut* to be the sum of the capacities of all edges in the cut, then it seems clear that the value of a flow can never exceed the capacity of any cut, and therefore that the *maximum* value of a flow cannot exceed the *minimum* capacity of any cut.

The main result of this section is the 'max-flow min-cut' theorem of Ford and Fulkerson, which we state as

Theorem 3.4.1. *The maximum possible value of any flow in a network is equal to the minimum capacity of any cut in that network.*

Proof: We will first do a little computation to show that the value of a flow can never exceed the capacity of a cut. Second, we will show that when the Ford-Fulkerson algorithm terminates because it has been unable to label the sink, then at that moment there is a cut in the network whose edges are saturated with flow, *i.e.*, such that the flow in each edge of the cut is equal to the capacity of that edge.

Let U and V be two (not necessarily disjoint) sets of vertices of the network $\mathbf{X}$, and let f be a flow function for $\mathbf{X}$. By $f(U,V)$ we mean the sum of the values of the flow function along all edges whose initial vertex lies in U and whose terminal vertex lies in V. Similarly, by $cap(U,V)$ we mean the sum of the capacities of all of those edges. Finally, by *the net flow out of* U we mean $f(U,\overline{U}) - f(\overline{U},U)$.

Lemma 3.4.1. *Let f be a flow of value Q in a network $\mathbf{X}$, and let $(W,\overline{W})$ be a cut in $\mathbf{X}$. Then*

$$Q = f(W,\overline{W}) - f(\overline{W},W) \leq cap(W,\overline{W}). \qquad (3.4.1)$$

Proof of lemma: The net flow out of s is Q. The net flow out of any other vertex $w \in W$ is 0. Hence, if $V(\mathbf{X})$ denotes the vertex set of the network $\mathbf{X}$, we obtain

$$\begin{aligned}
Q &= \sum_{w \in W} \{f(w,V(\mathbf{X})) - f(V(\mathbf{X}),w)\} \\
&= f(W,V(\mathbf{X})) - f(V(\mathbf{X}),W) \\
&= f(W,W \cup \overline{W}) - f(W \cup \overline{W},W) \\
&= f(W,W) + f(W,\overline{W}) - f(W,W) - f(\overline{W},W) \\
&= f(W,\overline{W}) - f(\overline{W},W).
\end{aligned}$$

This proves the '=' part of (3.4.1), and the '≤' part is obvious, completing the proof of lemma 3.4.1. ∎

We now know that the maximum value of the flow in a network cannot exceed the minimum of the capacities of the cuts in the network.

To complete the proof of the theorem we will show that a flow of maximum value, which surely exists, must saturate the edges of some cut.

116

Hence, let f be a flow in X of maximum value, and call procedure *labelandscan*$(\mathbf{X}, f, whyhalt)$. Let W be the set of vertices of $\mathbf{X}$ that have been labeled when the algorithm terminates. Clearly $s \in W$. Equally clearly, $t \notin W$, for suppose the contrary. Then we would have termination with '*whyhalt*' = 'it's time to augment,' and if we were then to call procedure *augmentflow* we would find a flow of higher value, contradicting the assumed maximality of f.

Since $s \in W$ and $t \notin W$, the set W defines a cut $(W, \overline{W})$.

We claim that every edge of the cut $(W, \overline{W})$ is saturated. Indeed, if (x, y) is in the cut, $x \in W$, $y \notin W$, then edge (x, y) is saturated, else y would have been labeled when we were scanning x and we would have $y \in W$, a contradiction. Similarly, if (y, x) is an edge where $y \in \overline{W}$ and $x \in W$, then the flow $f(y, x) = 0$, else again y would have been labeled when we were scanning x, another contradiction.

Therefore, every edge from W to $\overline{W}$ is carrying as much flow as its capacity permits, and every edge from $\overline{W}$ to W is carrying no flow at all. Hence the sign of equality holds in (3.4.1), the value of the flow is equal to the capacity of the cut $(W, \overline{W})$, and the proof of theorem 3.4.1 is finished. ∎

3.5 The complexity of the Ford-Fulkerson algorithm

The algorithm of Ford and Fulkerson terminates if and when it arrives at a stage where the sink is not labeled but no more vertices can be labeled. If at that time we let W be the set of vertices that have been labeled, then we have seen that $(W, \overline{W})$ is a minimum cut of the network, and the present value of the flow is the desired maximum for the network.

The question now is, how long does it take to arrive at that stage, and indeed, is it guaranteed that we will *ever* get there? We are asking if the algorithm is *finite*, surely the most primitive complexity question imaginable.

First consider the case where every edge of the given network $\mathbf{X}$ has *integer* capacity. Then during the labelling and flow augmentation algorithms, various additions and subtractions are done, but there is no way that any nonintegral flows can be produced.

It follows that the augmented flow is still integral. The *value* of the flow therefore increases by an integer amount during each augmentation. On the other hand if, say, C^* denotes the combined capacity of all edges that are outbound from the source, then it is eminently clear that the value of the flow can never exceed C^*. Since the value of the flow increases by at least 1 unit per augmentation, we see that no more than C^* flow augmentations will be needed before a maximum flow is reached. This yields

Theorem 3.5.1. *In a network with integer capacities on all edges, the Ford-Fulkerson algorithm terminates after a finite number of steps with a flow of maximum value.*

This is good news and bad news. The good news is that the algorithm is finite. The bad news is that the complexity estimate that we have proved depends not only on the numbers of edges and vertices in **X**, but on the edge capacities. If the bound C^* represents the true behavior of the algorithm, rather than some weakness in our analysis of the algorithm, then even on very small networks it will be possible to assign edge capacities so that the algorithm takes a very long time to run.

And it *is* possible to do that.

We will show below an example due to Ford and Fulkerson in which the situation is even worse than the one envisaged above: not only will the algorithm take a very long time to run; it won't converge at all!

Consider the network **X** that is shown in Fig. 3.5.1. It has 10 vertices s, t, $x_1, \ldots, x_4$, $y_1, \ldots, y_4$. There are directed edges (x_i, x_j) $\forall i \neq j$, (x_i, y_j) $\forall i, j$, (y_i, y_j) $\forall i \neq j$, (y_i, x_j) $\forall i, j$, (s, x_i) $\forall i$, and (y_j, t) $\forall j$.

In this network, the four edges $A_i = (x_i, y_i)$ $(i = 1, 4)$ will be called the *special edges*.

Next we will give the capacities of the edges of **X**. Write $r = (-1 + \sqrt{5})/2$, and let

$$S = (3 + \sqrt{5})/2 = \sum_{n=0}^{\infty} r^n.$$

Then to every edge of **X** except the four special edges we assign the capacity S. The special edges A_1, A_2, A_3, A_4 are given capacities $1, r, r^2, r^3$, respectively (you can see that this is going to be interesting).

Suppose, for our first augmentation step, we find the flow augmenting path $s \to x_1 \to y_1 \to t$, and that we augment the flow by 1 unit along that

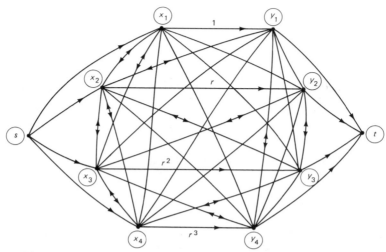

Fig. 3.5.1: How to give the algorithm a hard time

path. The four special edges will then have *residual capacities* (excesses of capacity over flow) of $0, r, r^2, r^3$, respectively.

Inductively, suppose we have arrived at a stage of the algorithm where the four special edges, taken in some rearrangement A'_1, A'_2, A'_3, A'_4, have residual capacities $0, r^n, r^{n+1}, r^{n+1}$. We will now show that the algorithm might next do two flow augmentation steps the net result of which would be that the inductive state of affairs would again hold, with n replaced by $n + 1$.

Indeed, choose the flow augmenting path

$$s \to x'_2 \to y'_2 \to x'_3 \to y'_3 \to t.$$

The only special edges that are on this path are A'_2 and A'_3. Augment the flow along this path by r^{n+1} units (the maximum possible amount).

Next, choose the flow augmenting path

$$s \to x'_2 \to y'_2 \to y'_1 \to x'_1 \to y'_3 \to x'_3 \to y'_4 \to t.$$

Notice that with respect to this path the special edges A'_1 and A'_3 are incoherently directed. Augment the flow along this path by r^{n+2} units, once more the largest possible amount.

The reader may now verify that the residual capacities of the four special edges are $r^{n+2}, 0, r^{n+2}, r^{n+1}$. In the course of doing this verification

119

it will be handy to use the fact that

$$r^{n+2} = r^n - r^{n+1} \qquad (\forall n \geq 0).$$

These two augmentation steps together have increased the flow value by $r^{n+1} + r^{n+2} = r^n$ units. Hence the flow in an edge will never exceed S units.

The algorithm converges to a flow of value S. Now comes the bad news: the maximum flow in this network has the value $4S$ (find it!).

Hence, for this network

(a) the algorithm does not halt after finitely many steps even though the edge capacities are finite and

(b) the sequence of flow values converges to a number that is not the maximum flow in the network.

The irrational capacities on the edges may at first seem to make this example seem 'cooked up.' But the implication is that even with a network whose edge capacities are all integers, the algorithm might take a very long time to run.

Motivated by the importance and beauty of the theory of network flows, and by the unsatisfactory time complexity of the original algorithm, many researchers have attacked the question of finding an algorithm whose success is guaranteed within a time bound that is independent of the edge capacities, and depends only on the size of the network.

We turn now to the consideration of one of the main ideas on which further progress has depended, that of *layering* a network with respect to a flow function. This idea has triggered a whole series of improved algorithms. Following the discussion of layering we will give a description of one of the algorithms, the MPM algorithm, that uses layered networks and guarantees fast operation.

3.6 Layered networks

Layering a network is a technique that has the effect of replacing a single max-flow problem by several problems, each a good deal easier than the original. More precisely, in a network with V vertices we will find that we can solve a max-flow problem by solving at most V slightly different problems, each on a layered network. We will then discuss an $O(V^2)$ method for solving each such problem on a layered network, and the result will be an $O(V^3)$ algorithm for the original network flow problem.

Now we will discuss how to *layer a network with respect to a given flow function*. The purpose of the italics is to emphasize the fact that one does not just 'layer a network.' Instead, there is given a network $\mathbf{X}$ and a flow function f for that network, and together they induce a layered network $\mathbf{Y} = \mathbf{Y}(\mathbf{X}, f)$, as follows.

First let us say that an edge e of $\mathbf{X}$ is *helpful from u to v* if either e is directed from u to v and $f(e)$ is below capacity or e is directed from v to u and the flow $f(e)$ is positive.

Next we will describe the layered network $\mathbf{Y}$. Recall that in order to describe a network one must describe the vertices of the network, the directed edges, give the capacities of those edges, and designate the source and the sink. The network $\mathbf{Y}$ will be constructed one layer at a time from the vertices of $\mathbf{X}$, using the flow f as a guide. For each layer, we will say which vertices of $\mathbf{X}$ go into that layer, then we will say which vertices of the previous layer are connected to each vertex of the new layer. All of these edges will be directed from the earlier layer to the later one. Finally we will give the capacities of each of these new edges.

The 0th layer of $\mathbf{Y}$ consists only of the source s. The vertices that comprise layer 1 of $\mathbf{Y}$ will be every vertex v of $\mathbf{X}$ such that in $\mathbf{X}$ there is a helpful edge from s to v. We then draw an edge in $\mathbf{Y}$ directed *from s to v* for each such vertex v. We assign to that edge in $\mathbf{Y}$ a capacity $cap(s,v) - f(s,v) + f(v,s)$.

The set of all such v will be called layer 1 of $\mathbf{Y}$. Next we construct layer 2 of $\mathbf{Y}$. The vertex set of layer 2 consists of all vertices w that do not yet belong to any layer, and such that there is a helpful edge in $\mathbf{X}$ from some vertex v of layer 1 to w.

Next we draw the edges from layer 1 to layer 2: for each vertex v in layer 1 we draw a single edge in $\mathbf{Y}$ directed from v to every vertex w in layer 2 for which there is a helpful edge in $\mathbf{X}$ from v to w.

Note that the edge always goes *from v to w* regardless of the direction of the helpful edge in $\mathbf{X}$. Note also that in contrast to the Ford-Fulkerson algorithm, even after an edge has been drawn from v to w in $\mathbf{Y}$, additional edges may be drawn to the same w from other vertices v', v'' in layer 1.

Assign capacities to the edges from layer 1 to layer 2 in the same way as described above, that is, the capacity in $\mathbf{Y}$ of the edge from v to w is $cap(v, w) - f(v, w) + f(w, v)$. This latter quantity is, of course, the total residual (unused) flow-carrying capacity of the edges in both directions between v and w.

The layering continues until we reach a layer L such that there is a helpful edge from some vertex of layer L to the sink t, or else until no additional layers can be created (to say that no more layers can be created is to say that among the vertices that haven't yet been included in the layered network that we are building, there aren't any that are adjacent to a vertex that is in the layered network, by a helpful edge).

In the former case, we then create a layer $L + 1$ that consists solely of the sink t, we connect t by edges directed from the appropriate vertices of layer L, assign capacities to those edges, and the layering process is complete. Observe that not all vertices of $\mathbf{X}$ need appear in $\mathbf{Y}$.

In the latter case, where no additional layers can be created but the sink hasn't been reached, the present flow function f in $\mathbf{X}$ is maximum, and the network flow problem in $\mathbf{X}$ has been solved.

Here is a formal statement of the procedure for layering a given network $\mathbf{X}$ with respect to a given flow function f in $\mathbf{X}$. Input are the network $\mathbf{X}$ and the present flow function f in that network. Output are the layered network $\mathbf{Y}$, and a logical variable *maxflow* that will be *True*, on output, if the flow is at a maximum value, *False* otherwise.

procedure *layer* (**X**, *f*, **Y**, *maxflow*);

{forms the layered network *Y* with respect to the flow *f* in **X** }

{*maxflow* will be '*True*' if the input flow *f* already has the
  maximum possible value for the network, else it will be '*False*'}

L:= 0; *layer*(*L*) := {*source*}; *maxflow* := *false*;

 repeat

  layer(*L* + 1) := $\emptyset$;

  for each vertex *u* in *layer*(*L*) **do**

    for each vertex *v* such that {layer(*v*) = *L* + 1 or *v* is

      not in any layer} **do**

    q := *cap*(*u*, *v*) − *f*(*u*, *v*) + *f*(*v*, *u*);

    if *q* > 0 **then do**

      draw edge *u* → *v* in **Y**;

      assign capacity *q* to that edge;

      assign vertex *v* to *layer*(*L* + 1);

  L := *L* + 1

  if *layer*(*L*) is empty then exit with *maxflow* := *true*;

 until *sink* is in *layer*(*L*);

 delete from *layer*(*L*) of *Y* all vertices other than *sink*,

  and remove their incident edges from *Y*

end.{*layer*}

In Fig. 3.6.1 we show the typical appearance of a layered network. In contrast to a general network, in a layered network every path from the source to some fixed vertex *v* has the same number of edges in it (the number of the layer of *v*), and all edges on such a path are directed the same way, from the source towards *v*. These properties of layered networks are very friendly indeed, and make them much easier to deal with than general networks.

In Fig. 3.6.2 we show specifically the layered network that results from the network of Fig. 3.1.2 with the flow shown therein.

The next question is this: exactly what problem would we like to solve on the layered network **Y**, and what is the relationship of that problem to the original network flow problem in the original network **X**?

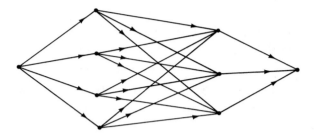

Fig. 3.6.1: A general layered network

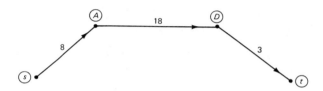

Fig. 3.6.2: A layering of the network in Fig. 3.1.2

The answer is that in the layered network **Y** we are looking for a *blocking flow g*. By a blocking flow we mean a flow function *g* in **Y** such that *every path from source to sink in* **Y** *has at least one saturated edge.*

This immediately raises two questions: (a) what can we do if we find a blocking flow in **Y**? (b) how can we find a blocking flow in **Y**? The remainder of this section will be devoted to answering (a). In the next section we will give an elegant answer to (b).

Suppose that we have somehow found a blocking flow function, *g*, in **Y**. What we do with it is that we use it to augment the flow function *f* in **X**, as follows.

procedure $augment(f, \mathbf{X}; g, \mathbf{Y})$;
{augment flow f in $\mathbf{X}$ by using a blocking flow g
    in the corresponding layered network $\mathbf{Y}$}
    for every edge $e : u \to v$ of the layered network $\mathbf{Y}$, do
        increase the flow f in the edge $u \to v$ of the
          network $\mathbf{X}$ by the amount
$$min\{g(e), cap(u \to v) - f(u \to v)\};$$
      if not all of $g(e)$ has been used
          then decrease the flow in edge $v \to u$ by
              the unused portion of $g(e)$
end.{*augment*}

After augmenting the flow in the original network $\mathbf{X}$, what then? We construct a new layered network, from $\mathbf{X}$ and the newly augmented flow function f on $\mathbf{X}$.

The various activities that are now being described may sound like some kind of thinly disguised repackaging of the Ford-Fulkerson algorithm, but they aren't just that, because here is what can be proved to happen:

First, if we start with zero flow in $\mathbf{X}$, make the layered network $\mathbf{Y}$, find a blocking flow in $\mathbf{Y}$, augment the flow in $\mathbf{X}$, make a new layered network $\mathbf{Y}$, find a blocking flow, etc. etc., then *after at most V phases* ('phase' = layer + block + augment) we will have found the maximum flow in $\mathbf{X}$ and the process will halt.

Second, each phase can be done very rapidly. The MPM algorithm, to be discussed in section 3.7, finds a blocking flow in a layered network in time $O(V^2)$.

By the *height* of a layered network $\mathbf{Y}$ we will mean the number of edges in any path from source to sink. The network of Fig. 3.6.1 has height 3. Let's now show

Theorem 3.6.1. *The heights of the layered networks that occur in the consecutive phases of the solution of a network flow problem form a strictly increasing sequence of positive integers. Hence, for a network X with V vertices, there can be at most V phases before a maximum flow is found.*

Let $\mathbf{Y}(p)$ denote the layered network that is constructed at the pth phase of the computation and let $H(p)$ denote the height of $\mathbf{Y}(p)$. We will first prove

Lemma 3.6.1. *If*

$$v_0 \to v_1 \to v_2 \to \cdots \to v_m \quad (v_0 = source)$$

is a path in $\mathbf{Y}(p+1)$, and if every vertex v_i $(i = 1, m)$ of that path also appears in $\mathbf{Y}(p)$, then for every $a = 0, m$ it is true that if vertex v_a was in layer b of $\mathbf{Y}(p)$ then $a \geq b$.

Proof of lemma: The result is clearly true for $a = 0$. Suppose it is true for $v_0, v_1, \ldots, v_a$, and suppose v_{a+1} was in layer c of network $\mathbf{Y}(p)$. We will show that $a + 1 \geq c$. Indeed, if not then $c > a + 1$. Since v_a, by induction, was in a layer $\leq a$, it follows that the edge

$$e^* : v_a \to v_{a+1}$$

was not present in network $\mathbf{Y}(p)$ since its two endpoints were not in two consecutive layers. Hence the flow in $\mathbf{Y}$ between v_a and v_{a+1} could not have been affected by the augmentation procedure of phase p. But edge e^* is in $\mathbf{Y}(p+1)$. Therefore it represented an edge of $\mathbf{Y}$ that was helpful from v_a to v_{a+1} at the beginning of phase $p + 1$, was unaffected by phase p, but was not helpful at the beginning of phase p. This contradiction establishes the lemma. ■

Now we will prove the theorem. Let

$$s \to v_1 \to v_2 \to \cdots \to v_{H(p+1)-1} \to t$$

be a path from source to sink in $\mathbf{Y}(p+1)$.

Consider first the case where every vertex of the path also lies in $\mathbf{Y}(p)$, and apply the lemma to $v_m = t$ $(m = H(p+1)), a = m$. We conclude at once that $H(p+1) \geq H(p)$. Now we want to exclude the '=' sign. If $H(p+1) = H(p)$ then the entire path is in $\mathbf{Y}(p)$ and in $\mathbf{Y}(p+1)$, and so all of the edges in $\mathbf{Y}$ that the edges of the path represent were helpful both before and after the augmentation step of phase p, contradicting the fact that the blocking flow that was used for the augmentation saturated some edge of the chosen path. The theorem is now proved for the case where the path had all of its vertices in $\mathbf{Y}(p)$ also.

Now suppose that this was not the case. Let $e^* : v_a \to v_{a+1}$ be the first edge of the path whose terminal vertex v_{a+1} was not in $\mathbf{Y}(p)$. Then the corresponding edge(s) of $\mathbf{Y}$ was unaffected by the augmentation in phase p. It was helpful from v_a to v_{a+1} at the beginning of phase $p+1$ because $e^* \in \mathbf{Y}(p+1)$ and it was unaffected by phase p, yet $e^* \notin \mathbf{Y}(p)$. The only possibility is that vertex v_{a+1} would have entered into $\mathbf{Y}(p)$ in the layer $H(p)$ that contains the sink, but that layer is special, and contains only t. Hence, if v_a was in layer b of $\mathbf{Y}(p)$, then $b+1 = H(p)$. By the lemma once more, $a \geq b$, so $a+1 \geq b+1 = H(p)$, and therefore $H(p+1) > H(p)$, completing the proof of theorem 3.6.1. ■

To summarize, if we want to find a maximum flow in a given network $\mathbf{Y}$ by the method of layered networks, we carry out

> procedure $maxflow$ $(\mathbf{X}, \mathbf{Y}, f)$;
> set the flow function f to zero on all edges of $\mathbf{Y}$;
> **repeat**
> > (i) construct the layered network $\mathbf{Y} = \mathbf{Y}(\mathbf{X}, f)$ if possible, else exit with flow at maximum value;
> > (ii) find a blocking flow g in $\mathbf{Y}$;
> > (iii) augment the flow f in $\mathbf{Y}$ with the blocking flow g, by calling procedure $augment$ above
> **until** exit occurs in (i) above;
> end.$\{maxflow\}$

According to theorem 3.6.1, the procedure will repeat steps (i), (ii), (iii) at most V times because the height of the layered network increases each time around, and it certainly can never exceed V. The labor involved in step (i) is certainly $O(E)$, and so is the labor in step (iii). Hence if BFL denotes the labor involved in some method for finding a blocking flow in a layered network, then the whole network flow problem can be done in time $O(V \cdot (E + BFL))$.

The idea of layering networks is due to Dinic. Since his work was done, all efforts have been directed at the problem of reducing BFL as much as possible.

3.7 The MPM algorithm

Now we suppose that we are given a *layered* network **Y** and we want to find a blocking flow g for **Y**. The following ingenious suggestion is due to Malhotra, Pramodh-Kumar and Maheshwari.

Let v be some vertex of **Y**. The *in-potential* of v is the sum of the capacities of all edges directed into v, and the *outpotential* of v is the total capacity of all edges directed out from v. The *potential* of v is the smaller of these two.

(A) Find a vertex v of smallest potential, say P^*. Now we will push P^* more units of flow from source to sink, as follows.

(B) (Pushout) Take the edges that are outbound from v in some order, and saturate each one with flow, unless and until saturating one more would lift the total flow used over P^*. Then assign all remaining flow to the next outbound edge (not necessarily saturating it), so the total outflow from v becomes exactly P^*.

(C) Follow the flow to the next higher layer of **Y**. That is, for each vertex v' of the next layer, let $h(v')$ be the flow into v'. Now saturate all except possibly one outbound edge of v', to pass through v' the $h(v')$ units of flow. When all vertices v' in that layer have been done, repeat for the next layer, etc. We never find a vertex with insufficient capacity, in or out, to handle the flow that is thrust upon it, because we began by choosing a vertex of minimum potential.

(D) (Pullback) When all layers 'above' v have been done, then follow the flow to the next layer 'below' v. For each vertex v' of that layer, let $h(v')$ be the flow out of v' to v. Then saturate all except possibly one *incoming* edge of v', to pass through v' the $h(v')$ units of flow. When all v' in that layer have been done, proceed to the next layer below v, etc.

(E) (Update capacities) The flow function that has just been created in the layered network must now be stored somewhere. A convenient way to keep it is to carry out the augmentation procedure back in the network **X** at this time, thereby, in effect, 'storing' the contributions to the blocking flow in **Y** in the flow array for **X**. This can be done concurrently with the MPM algorithm as follows: Every time we increase the flow in some edge $u \to v$ of **Y** we do it by augmenting the flow from u to v in **X**, and then decreasing the capacity of edge $u \to v$ in **Y** by the same amount. In that way, the capacities of the edges in **Y** will always be the updated *residual* capacities, and the flow function f in $\mathbf{X}$ will always reflect the latest augmentation of the flow in $\mathbf{Y}$.

(F) (Prune) We have now pushed the original $h(v)$ units of flow through the whole layered network. We intend to repeat the operation on some other vertex v of minimum potential, but first we can prune off of the network some vertices and edges that are guaranteed never to be needed again.

The vertex v itself has either all incoming edges or all outgoing edges, or both, at zero residual capacities. Hence no more flow will ever be pushed through v. Therefore we can delete v from the network $\mathbf{Y}$ together with all of its incident edges, incoming or outgoing. Further, we can delete from $\mathbf{Y}$ all of the edges that were saturated by the flow pushing process just completed, *i.e.*, all edges that now have zero residual capacity.

Next, we may now find that some vertex w has had all of its incoming edges or all of its outgoing edges deleted. That vertex will never be used again, so delete it and any other incident edges it may still have. Continue the pruning process until only vertices remain that have nonzero potential. If the source and sink are still connected by some path, then repeat from (A) above.

Else, the algorithm halts. The blocking flow function g that we have just found is the following: if e is an edge of the input layered network $\mathbf{Y}$, then $g(e)$ is the sum of all of the flows that were pushed through edge e at all stages of the above algorithm.

It is obviously a blocking flow: since no path between s and t remains, every path must have had at least one of its edges saturated at some step of the algorithm. What is the complexity of this algorithm? Certainly we delete at least one vertex from the network at every pruning stage, because the vertex v that had minimum potential will surely have had either all of its incoming or its outgoing edges (or both) saturated.

It follows that the steps (A) – (E) can be executed at most V times before we halt with a blocking flow. The cost of saturating all edges that get saturated, since every edge has but one saturation to give to its network, is $O(E)$. The number of partial edge-saturation operations is at most two per vertex visited. For each minimal-potential vertex v we visit at most V other vertices, and we use at most V minimal-potential vertices altogether. So the partial edge-saturation operations cost $O(V^2)$ and the total edge saturations cost $O(E)$.

The operation of finding a vertex of minimum potential is 'free,' in the following sense. Initially we compute and store the in- and out- potentials of every vertex. Thereafter, each time the flow in some edge is increased, the outpotential of its initial vertex and the inpotential of its terminal vertex are reduced by the same amount. It follows that the cost of maintaining these arrays is linear in the number of vertices, V. Hence it affects only the constants implied by the 'big oh' symbols above, but not the orders of magnitude.

The total cost is therefore $O(V^2)$ for the complete MPM algorithm that finds a blocking flow in a layered network. Hence a maximum flow in a network can be found in $O(V^3)$ time, since at most V layered networks need to be looked at in order to find a maximum flow in the original network.

In contrast to the nasty example network of section 3.5, with its irrational edge capacities, that made the Ford-Fulkerson algorithm into an infinite process that converged to the wrong answer, the time bound $O(V^3)$ that we have just proved for the layered-network-MPM algorithm is totally independent of edge capacities.

3.8 Applications of network flow

We conclude this chapter by mentioning some applications of the network flow problem and algorithm.

Certainly, among these, one most often mentions first the problem of maximum matching in a bipartite graph. Consider a set of P people and a set of J jobs, such that not all of the people are capable of doing all of the jobs.

We construct a graph of $P + J$ vertices to represent this situation, as follows. Take P vertices to represent the people, J vertices to represent the jobs, and connect vertex p to vertex j by an undirected edge if person p can do job j. Such a graph is called *bipartite*. In general a graph G is bipartite if its vertices can be partitioned into two classes in such a way that no edges of G run between two vertices of the same class (see section 1.6).

In Fig. 3.8.1 below we show a graph that might result from a certain group of 8 people and 9 jobs.

The maximum matching problem is just this: assuming that each per-

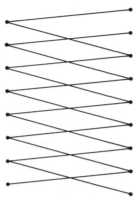

Fig. 3.8.1: Matching people to jobs

son can handle at most one of the jobs, and that each job needs only one person, assign people to jobs in such a way that the largest possible number of people are employed. In terms of the bipartite graph G, we want to *find a maximum number of edges, no two incident with the same vertex.*

To solve this problem by the method of network flows we construct a network **Y**. First we adjoin two new vertices s, t to the bipartite graph G. If we let P, J denote the two classes of vertices in the graph G, then we draw an edge from s to each $p \in P$ and an edge from each $j \in J$ to t. Each edge in the network is given capacity 1. The result for the graph of Fig. 3.8.1 is shown in Fig. 3.8.2.

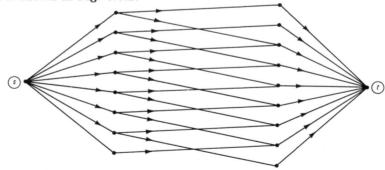

Fig. 3.8.2: The network for the matching problem

Consider a maximum integer-valued flow in this network, of value Q. Since each edge has capacity 1, Q edges of the type (s, p) each contain a unit of flow. Out of each vertex p that receives some of this flow there will come one unit of flow (since inflow equals outflow at such vertices), which will then cross to a vertex j of J. No such j will receive more than one

131

unit because at most one unit can leave it for the sink t. Hence the flow defines a matching of Q edges of the graph G. Conversely, any matching in G defines a flow, hence a maximum flow corresponds to a maximum matching. In Fig. 3.8.3 we show a maximum flow in the network of Fig. 3.8.2 and therefore a maximum matching in the graph of Fig. 3.8.1.

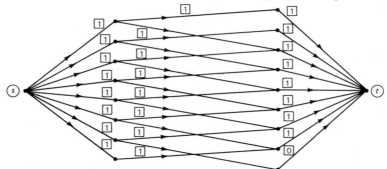

Fig. 3.8.3: A maximum flow

For a second application of network flow methods, consider an undirected graph G. The *edge-connectivity* of G is defined as the minimum number of edges of G whose removal would disconnect G. Certainly, for instance, if we remove all of the edges incident to a single vertex v, we will disconnect the graph. Hence the edge connectivity cannot exceed the minimum degree of vertices in the graph. However, the edge connectivity could be quite a lot smaller than the minimum degree, as the graph of Fig. 3.8.4 shows, in which the minimum degree is large, but the removal of just one edge will disconnect the graph.

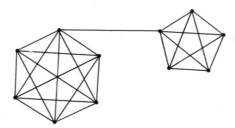

Fig. 3.8.4: Big degree, low connectivity

Finding the edge connectivity is quite an important combinatorial problem, and it is by no means obvious that network flow methods can be used on it, but they can, and here is how.

Given G, a graph of V vertices. We solve not just one but $V - 1$ network flow problems, one for each vertex $j = 2, \ldots, V$.

Fix such a vertex j. Then consider vertex 1 of G to be the source and vertex j to be the sink of a network $\mathbf{X}_j$. Replace each edge of G by two edges of $\mathbf{X}_j$, one in each direction, each with capacity 1. Now solve the network flow problem in X_j obtaining a maximum flow $Q(j)$. Then the smallest of the numbers $Q(j)$, for $j = 2, \ldots, V$ is the edge connectivity of G. We will not prove this here.*

As a final application of network flow we discuss the beautiful question of determining whether or not there is a matrix of 0's and 1's that has given row and column sums. For instance, is there a 6×8 matrix whose row sums are respectively $(5, 5, 4, 3, 5, 6)$ and whose column sums are $(3, 4, 4, 4, 3, 3, 4, 3)$? Of course, the phrase 'row sums' means the same thing as 'number of 1's in each row' since we have said that the entries are only 0 or 1.

Hence in general, let there be given a row-sum vector $(r_1, \ldots, r_m)$ and a column-sum vector $(s_1 \ldots, s_n)$. We ask if there exists an $m \times n$ matrix A of 0's and 1's that has exactly r_i 1's in the ith row and exactly s_j 1's in the jth column, for each $i = 1, \ldots, m$, $j = 1, \ldots, n$. The reader will no doubt have noticed that for such a matrix to exist it must surely be true that

$$r_1 + \cdots + r_m = s_1 + \cdots + s_n \qquad (3.8.1)$$

since each side counts the total number of 1's in the matrix. Hence we will suppose that (3.8.1) is true.

Now we will construct a network $\mathbf{Y}$ of $m + n + 2$ vertices, named s, $x_1, \ldots, x_m, y_1, \ldots, y_n, t$. There is an edge of capacity r_i drawn from the source s to vertex x_i, for each $i = 1, \ldots, m$ and an edge of capacity s_j drawn from vertex y_j to the sink t, for each $j = 1, \ldots, n$. Finally, there are mn edges of capacity 1 drawn from each x_i to each vertex y_j.

Next find a maximum flow in this network. Then, there is a 0-1 matrix with the given row and column sum vectors if and only if a maximum flow saturates every edge outbound from the source, that is, if and only if a maximum flow has value equal to the right or left side of equation (3.8.1).

---

* S. Even and R. E. Tarjan, Network flow and testing graph connectivity, *SIAM J. Computing*, **4** (1975), 507-518.

If such a flow exists then a matrix A of the desired kind is constructed by putting a_{ij} equal to the flow in the edge from x_i to y_j.

Exercises for section 3.8

1. Apply the max-flow min-cut theorem to the network that is constructed in order to solve the bipartite matching problem. Precisely what does a cut correspond to in this network? What does the theorem tell you about the matching problem?

2. Same as question 1 above, but applied to the problem of discovering whether or not there is a 0-1 matrix with a certain given set of row and column sums.

Bibliography

The standard reference for the network flow problem and its variants is

L. R. Ford and D. R. Fulkerson, *Flows in Networks*, Princeton University Press, Princeton, NJ, 1974.

The algorithm, the example of irrational capacities and lack of convergence to maximum flow, and many applications are discussed there. The chronology of accelerated algorithms is based on the following papers. The first algorithms with a time bound independent of edge capacities are in

J. Edmonds and R. M. Karp, Theoretical improvements in algorithmic efficiency for network flow problems, *JACM* **19**, 2 (1972), 248-264.

E. A. Dinic, Algorithm for solution of a problem of maximal flow in a network with power estimation, *Soviet Math. Dokl.*, **11** (1970), 1277-1280.

The paper of Dinic, above, also originated the idea of a layered network. Further accelerations of the network flow algorithms are found in the following.

A. V. Karzanov, Determining the maximal flow in a network by the method of preflows, *Soviet Math. Dokl.*, **15** (1974), 434-437.

B. V. Cherkassky, Algorithm of construction of maximal flow in networks with complexity of $O(V^2 E)$ operations, Akad. Nauk USSR, *Mathematical methods for the solution of economical problems* **7** (1977), 117-126.

The MPM algorithm, discussed in the text, is due to

V. M. Malhotra, M. Pramodh-Kumar and S. N. Maheshwari, An $O(V^3)$ algorithm for finding maximum flows in networks, *Information Processing Letters*, **7** (1978), 277-278.

Later algorithms depend on refined data structures that save fragments of partially constructed augmenting paths. These developments were initiated in

Z. Galil, A new algorithm for the maximal flow problem, *Proc.* 19th *IEEE Symposium on the Foundations of Computer Science*, Ann Arbor, October 1978, 231-245.

Andrew V. Goldberg and Robert E. Tarjan, A new approach to the Maximum Flow Problem, 1985, to appear.

A number of examples that show that the theoretical complexity estimates for the various algorithms cannot be improved are contained in

Z. Galil, On the theoretical efficiency of various network flow algorithms, IBM report RC7320, September 1978.

The proof given in the text, of theorem 3.6.1, leans heavily on the one in

Shimon Even, *Graph Algorithms*, Computer Science Press, Potomac, MD, 1979.

If edge capacities are all 0's and 1's, as in matching problems, then still faster algorithms can be given, as in

S. Even and R. E. Tarjan, Network flow and testing graph connectivity, *SIAM J. Computing*, **4** (1975), 507-518.

If every pair of vertices is to act, in turn, as source and sink, then considerable economies can be realized. See

R. E. Gomory and T. C. Hu, Multiterminal network flows, *SIAM Journal*, **9** (1961), 551-570.

Matching in general graphs is much harder than in bipartite graphs. The pioneering work is due to

J. Edmonds, Paths, trees and flowers, *Canadian J. Math.*, **17** (1965), 449-467.

Chapter 4: Algorithms in the Theory of Numbers

Number theory is the study of the properties of the positive integers. It is one of the oldest branches of mathematics, and one of the purest, so to speak. It has immense vitality, however, and we will see in this chapter and the next that parts of number theory are extremely relevant to current research in algorithms.

Part of the reason for this is that number theory enters into the analysis of algorithms, but that isn't the whole story.

Part of the reason is that many famous problems of number theory, when viewed from an algorithmic viewpoint (like, how do you decide whether or not a positive integer n is prime?) present extremely deep and attractive unsolved algorithmic problems. At least, they are unsolved if we regard the question as not just how to do these problems computationally, but how to do them as rapidly as possible.

But that's not the whole story either.

There are close connections between algorithmic problems in the theory of numbers, and problems in other fields, seemingly far removed from number theory. There is a unity between these seemingly diverse problems that enhances the already considerable beauty of any one of them. At least some of these connections will be apparent by the end of study of Chapter 5.

4.1 Preliminaries

We collect in this section a number of facts about the theory of numbers, for later reference.

If n and m are positive integers then to *divide* n by m is to find an integer $q \geq 0$ (the *quotient*) and an integer r (the *remainder*) such that $0 \leq r < m$ and $n = qm + r$.

If $r = 0$, we say that 'm divides n,' or 'm is a divisor of n,' and we write $m|n$. In any case the remainder r is also called 'n modulo m,' and we write $r = n \bmod m$. Thus $4 = 11 \bmod 7$, for instance.

If n has no divisors other than $m = n$ and $m = 1$, then n is *prime*, else n is *composite*. Every positive integer n can be factored into primes, uniquely apart from the order of the factors. Thus $120 = 2^3 \cdot 3 \cdot 5$, and in general we will write

$$n = p_1^{a_1} p_2^{a_2} \cdots p_l^{a_l} = \prod_{i=1}^{l} p_i^{a_i}. \qquad (4.1.1)$$

We will refer to (4.1.1) as the *canonical factorization* of n.

Many interesting and important properties of an integer n can be calculated from its canonical factorization. For instance, let $d(n)$ be the number of divisors of the integer n. The divisors of 6 are 1, 2, 3, 6, so $d(6) = 4$.

Can we find a formula for $d(n)$? A small example may help to clarify the method. Since $120 = 2^3 \cdot 3 \cdot 5$, a divisor of 120 must be of the form $m = 2^a 3^b 5^c$, in which a can have the values 0,1,2,3, b can be 0 or 1, and c can be 0 or 1. Thus there are 4 choices for a, 2 for b and 2 for c, so there are 16 divisors of 120.

In general, the integer n in (4.1.1) has exactly

$$d(n) = (1 + a_1)(1 + a_2) \cdots (1 + a_l) \qquad (4.1.2)$$

divisors.

If m and n are nonnegative integers then their *greatest common divisor*, written $gcd(n, m)$, is the integer g that

(a) divides both m and n and

(b) is divisible by every other common divisor of m and n.

Thus $gcd(12, 8) = 4$, $gcd(42, 33) = 3$, etc. If $gcd(n, m) = 1$ then we say that n and m are *relatively prime*. Thus 27 and 125 are relatively prime (even though neither of them is prime).

If $n > 0$ is given, then $\phi(n)$ will denote the number of positive integers m such that $m \le n$ and $gcd(n, m) = 1$. Thus $\phi(6) = 2$, because there are only two positive integers ≤ 6 that are relatively prime to 6 (namely 1 and 5). $\phi(n)$ is called the Euler ϕ-function, or the Euler *totient* function.

Let's find a formula that expresses $\phi(n)$ in terms of the canonical factorization (4.1.1) of n.

We want to count the positive integers m for which $m \le n$, and m is not divisible by any of the primes p_i that appear in (4.1.1). There are n possibilities for such an integer m. Of these we throw away n/p_1 of them because they are divisible by p_1. Then we discard n/p_2 multiples of p_2, etc. This leaves us with

$$n - n/p_1 - n/p_2 - \cdots - n/p_l \qquad (4.1.3)$$

possible m's.

But we have thrown away too much. An integer m that is a multiple of both p_1 and p_2 has been discarded at least twice. So let's correct these errors by adding

$$n/(p_1 p_2) + n/(p_1 p_3) + \cdots + n/(p_1 p_l) + \cdots + n/(p_{l-1} p_l)$$

to (4.1.3).

The reader will have noticed that we added back too much, because an integer that is divisible by $p_1 p_2 p_3$, for instance, would have been re-entered at least twice. The 'bottom line' of counting too much, then too little, then too much, etc. is the messy formula

$$\begin{aligned}
\phi(n) =\; & n - n/p_1 - n/p_2 - \cdots - n/p_l + n/(p_1 p_2) + \cdots + n/(p_{l-1} p_l) \\
& - n/(p_1 p_2 p_3) - \cdots - n/(p_{l-2} p_{l-1} p_l) \\
& + \cdots + (-1)^l n/(p_1 p_2 \cdots p_l).
\end{aligned}$$
$$(4.1.4)$$

Fortunately (4.1.4) is identical with the much simpler expression

$$\phi(n) = n(1 - 1/p_1)(1 - 1/p_2) \cdots (1 - 1/p_l) \qquad (4.1.5)$$

which the reader can check by beginning with (4.1.5) and expanding the product.

To calculate $\phi(120)$, for example, we first find the canonical factorization $120 = 2^3 \cdot 3 \cdot 5$. Then we apply (4.1.5) to get

$$\phi(120) = 120(1 - 1/2)(1 - 1/3)(1 - 1/5)$$
$$= 32.$$

Thus, among the integers $1, 2, \ldots, 120$, there are exactly 32 that are relatively prime to 120.

Exercises for section 4.1

1. Find a formula for the sum of the divisors of an integer n, expressed in terms of its prime divisors and their multiplicities.

2. How many positive integers are $\leq 10^{10}$ and have an odd number of divisors? Find a simple formula for the number of such integers that are $\leq n$.

3. If $\phi(n) = 2$ then what do you know about n?

4. For which n is $\phi(n)$ odd?

4.2 The greatest common divisor

Let m and n be two positive integers. Suppose we divide n by m, to obtain a quotient q and a remainder r, with, of course, $0 \leq r < m$. Then we have

$$n = qm + r. \tag{4.2.1}$$

If g is some integer that divides both n and m then obviously g divides r also. Thus every common divisor of n and m is a common divisor of m and r. Conversely, if g is a common divisor of m and r then (4.2.1) shows that g divides n too.

It follows that $gcd(n, m) = gcd(m, r)$. If $r = 0$ then $n = qm$, and clearly, $gcd(n, m) = m$.

If we use the customary abbreviation '$n \bmod m$' for r, the remainder in the division of n by m, then what we have shown is that

$$gcd(n, m) = gcd(m, n \bmod m).$$

This leads to the following recursive procedure for computing the g.c.d.

```
function gcd(n, m);
{finds gcd of given nonnegative integers n and m}
    if m = 0  then gcd := n  else gcd := gcd(m, n mod m)
end.
```

The above is the famous 'Euclidean algorithm' for the g.c.d. It is one of the oldest algorithms known.

The reader is invited to write the Euclidean algorithm as a recursive program, and get it working on some computer. Use a recursive language, write the program more or less as above, and try it out with some large, healthy integers n and m.

The *gcd* program exhibits all of the symptoms of recursion. It calls itself with smaller values of its variable list. It begins with 'if trivialcase then do trivialthing' ($m = 0$), and this case is all-important because it's the only way the procedure can stop itself.

If, for example, we want the g.c.d. of 13 and 21, we call the program with $n = 13$ and $m = 21$, and it then recursively calls itself with the following arguments:

$$(21,13), \ (13,8), \ (8,5), \ (5,3), \ (3,2), \ (2,1), \ (1,0) \qquad (4.2.2)$$

When it arrives at a call in which the 'm' is 0, then the 'n,' namely 1 in this case, is the desired g.c.d.

What is the input to the problem? The two integers n, m whose g.c.d. we want are the input, and the number of bits that are needed to input those two integers is $\Theta(\log n) + \Theta(\log m)$, namely $\Theta(\log mn)$. Hence $c \log mn$ is the length of the input bit string. Now let's see how long the algorithm might run with an input string of that length.

To measure the running time of the algorithm we need first to choose a unit of cost or work. Let's agree that one unit of labor is the execution of a single '$a \bmod b$' operation. In this problem, an equivalent measure of cost would be the number of times the algorithm calls itself recursively. In the example (4.2.2) the cost was 7 units.

Lemma 4.2.1. *If $1 \le b \le a$ then $a \bmod b \le (a-1)/2$.*

Proof: Clearly $a \bmod b \le b - 1$. Further,

$$a \bmod b = a - \left\lfloor \frac{a}{b} \right\rfloor b$$
$$\le a - b.$$

Thus $a \bmod b \le \min(a - b, b - 1)$. Now we distinguish two cases.

First suppose $b \le (a+1)/2$. Then $b - 1 \le a - b$ and so

$$a \bmod b \le b - 1$$
$$\le \frac{a+1}{2} - 1$$
$$= \frac{a-1}{2}$$

in this case.

Next, suppose $b > (a+1)/2$. Then $a - b \leq b - 1$ and

$$a \bmod b \leq a - b < a - \frac{a+1}{2} = \frac{a-1}{2}$$

so the result holds in either case. ■

Theorem 4.2.1. *(A worst-case complexity bound for the Euclidean algorithm) Given two positive integers a, b. The Euclidean algorithm will find their greatest common divisor after a cost of at most $\lfloor 2 \log_2 M \rfloor + 1$ integer divisions, where $M = \max(a, b)$.*

Before we prove the theorem, let's return to the example $(a, b) = (13, 21)$ of the display (4.2.2). In that case $M = 21$ and $2 \log_2 M + 1 = 9.78\ldots$. The theorem asserts that the g.c.d. will be found after at most 9 operations. In fact it was found after 7 operations in that case.

Proof of theorem: Suppose first that $a \geq b$. The algorithm generates a sequence $a_0, a_1, \ldots$ where $a_0 = a, a_1 = b$, and

$$a_{j+1} = a_{j-1} \bmod a_j \qquad (j \geq 1).$$

By lemma 4.2.1,

$$a_{j+1} \leq \frac{a_{j-1} - 1}{2}$$
$$\leq \frac{a_{j-1}}{2}.$$

Then, by induction on j it follows that

$$a_{2j} \leq \frac{a_0}{2^j} \qquad (j \geq 0)$$

$$a_{2j+1} \leq \frac{a_1}{2^j} \qquad (j \geq 0)$$

and so,

$$a_r \leq 2^{-\lfloor r/2 \rfloor} M \qquad (r = 0, 1, 2, \ldots).$$

Obviously the algorithm has terminated if $a_r < 1$, and this will have happened when r is large enough so that $2^{-\lfloor r/2 \rfloor} M < 1$, *i.e.*, if $r > 2 \log_2 M$. If $a < b$ then after 1 operation we will be in the case '$a \geq b$' that we have just discussed, and the proof is complete. ■

The upper bound in the statement of theorem 4.2.1 can be visualized as follows. The number $\log_2 M$ is almost exactly the number of bits in the binary representation of M (what is 'exactly' that number of bits?). Theorem 4.2.1 therefore asserts that we can find the g.c.d. of two integers in a number of operations that is at most a linear function of the number of bits that it takes to represent the two numbers. In brief, we might say that 'Time = O(bits),' in the case of Euclid's algorithm.

Exercises for section 4.2

1. Write a nonrecursive program, in Basic or Fortran, for the g.c.d. Write a recursive program, in Pascal or a recursive language of your choice, for the g.c.d.

2. Choose 1000 pairs of integers (n, m), at random between 1 and 1000. For each pair, compute the g.c.d. using a recursive program and a nonrecursive program.

   (a) Compare the execution times of the two programs.

   (b) There is a theorem to the effect that the probability that two random integers have g.c.d. $= 1$ is $6/\pi^2$. What, precisely, do you think that this theorem means by 'the probability that ...'? What percentage of the 1000 pairs that you chose had g.c.d. $= 1$? Compare your observed percentage with $100 \cdot (6/\pi^2)$.

3. Find out when Euclid lived, and with exactly what words he described his algorithm.

4. Write a program that will light up a pixel in row m and column n of your CRT display if and only if $gcd(m, n) = 1$. Run the program with enough values of m and n to fill your screen. If you see any interesting visual patterns, try to explain them mathematically.

5. Show that if m and n have a total of B bits, then Euclid's algorithm will not need more than $2B + 3$ operations before reaching termination.

6. Suppose we have two positive integers m, n, and we have factored them completely into primes, in the form

$$m = \prod p_i^{a_i}; \qquad n = \prod q_i^{b_i}.$$

How would you calculate $gcd(m, n)$ from the above information? How would you calculate the least common multiple (*lcm*) of m and n from the above information? Prove that $gcd(m,n) = mn/lcm(m,n)$.

7. Calculate $gcd(102131, 56129)$ in two ways: use the method of exercise 6 above, then use the Euclidean algorithm. In each case count the total number of arithmetic operations that you had to do to get the answer.

8. Let F_n be the nth Fibonacci number. How many operations will be needed to compute $gcd(F_n, F_{n-1})$ by the Euclidean algorithm? What is $gcd(F_n, F_{n-1})$?

4.3 The extended Euclidean algorithm

Again suppose n, m are two positive integers whose g.c.d. is g. Then we can always write g in the form

$$g = tn + um \qquad (4.3.1)$$

where t and u are integers. For instance, $gcd(14, 11) = 1$, so we can write $1 = 14t + 11u$ for integers t, u. Can you spot integers t, u that will work? One pair that does the job is $(4, -5)$, and there are others (can you find all of them?).

The extended Euclidean algorithm finds not only the g.c.d. of n and m, it also finds a pair of integers t, u that satisfy (4.3.1). One 'application' of the extended algorithm is that we will obtain an inductive proof of the *existence* of t, u, that is not immediately obvious from (4.3.1) (see exercise 1 below). While this hardly rates as a 'practical' application, it represents a very important feature of recursive algorithms. We might say, rather generally, that the following items go hand-in-hand:

Recursive algorithms

Inductive proofs

Complexity analyses by recurrence formulas

If we have a recursive algorithm, then it is natural to prove the validity of the algorithm by mathematical induction. Conversely, inductive proofs of theorems often (not always, alas!) yield recursive algorithms for the construction of the objects that are being studied. The complexity analysis of a recursive algorithm will use recurrence formulas, in a natural way. We saw that already in the analysis that proved theorem 4.2.1.

Now let's discuss the extended algorithm. Input to it will be two integers n and m. Output from it will be $g = gcd(n, m)$ *and* two integers t and u for which (4.3.1) is true.

A single step of the original Euclidean algorithm took us from the problem of finding $gcd(n, m)$ to $gcd(m, n \bmod m)$. Suppose, inductively, that we not only know $g = gcd(m, n \bmod m)$ but we also know the coefficients t', u' for the equation

$$g = t'm + u'(n \bmod m). \tag{4.3.2}$$

Can we get out, at the next step, the corresponding coefficients t, u for (4.3.1)? Indeed we can, by substituting in (4.3.2) the fact that

$$n \bmod m = n - \left\lfloor \frac{n}{m} \right\rfloor m \tag{4.3.3}$$

we find that

$$
\begin{aligned}
g &= t'm + u'(n - \left\lfloor \frac{n}{m} \right\rfloor m) \\
&= u'n + (t' - u' \left\lfloor \frac{n}{m} \right\rfloor)m.
\end{aligned}
\tag{4.3.4}
$$

Hence the rule by which t', u' for equation (4.3.2) transform into t, u for equation (4.3.1) is that

$$
\begin{aligned}
t &= u' \\
u &= t' - \left\lfloor \frac{n}{m} \right\rfloor u'.
\end{aligned}
\tag{4.3.5}
$$

We can now formulate recursively the extended Euclidean algorithm.

```
procedure gcdext(n, m, g, t, u);
{computes g.c.d. of n and m, and finds
    integers t, u that satisfy (4.3.1)}
  if m = 0  then
        g := n; t := 1; u := 0
        else
        gcdext(m, n mod m, g, t, u);
        s := u;
        u := t − ⌊n/m⌋ u;
        t := s
  end.{gcdext}
```

It is quite easy to use the algorithm above to make a proof of the main mathematical result of this section (see exercise 1), which is

Theorem 4.3.1. *Let m and n be given integers, and let g be their greatest common divisor. Then there exist integers t, u such that $g = tm + un$.* ■

An immediate consequence of the algorithm and the theorem is the fact that finding inverses modulo a given integer is an easy computational problem. We will need to refer to that fact in the sequel, so we state it as

Corollary 4.3.1. *Let m and n be given positive integers, and let g be their g.c.d. Then m has a multiplicative inverse modulo n if and only if $g = 1$. In that case, the inverse can be computed in polynomial time.*

Proof: By the extended Euclidean algorithm we can find, in linear time, integers t and u such that $g = tm + un$. But this last equation says that $tm \equiv g \pmod{n}$. If $g = 1$ then it is obvious that t is the inverse mod n of m. If $g > 1$ then there exists no t such that $tm \equiv 1 \pmod{n}$ since $tm = 1 + rn$ implies that the g.c.d. of m and n is 1. ■

We will now trace the execution of *gcdext* if it is called with $(n, m) = (14, 11)$. The routine first replaces (14,11) by (11,3) and calls itself. Then it calls itself successively with (3,2), (2,1) and (1,0). When it executes with $(n, m) = (1, 0)$ it encounters the 'if $m = 0$' statement, so it sets $g := 1, t := 1, u := 0$.

Now it can complete the execution of the call with $(n, m) = (2, 1)$, which has so far been pending. To do this it sets

$$u := t - \lfloor n/m \rfloor u = 1$$

$$t := 0.$$

The call with $(n, m) = (2, 1)$ is now complete. The call to the routine with $(n, m) = (3, 2)$ has been in limbo until just this moment. Now that the (2,1) call is finished, the (3,2) call executes and finds

$$u := 0 - \lfloor 3/2 \rfloor 1 = 1$$

$$t := 1.$$

The call to the routine with $(n, m) = (11, 3)$ has so far been languishing, but its turn has come. It computes

$$u := 1 - \lfloor 11/3 \rfloor (-1) = 4$$

$$t := -1.$$

Finally, the original call to *gcdext* from the user, with $(n, m) = (14, 11)$, can be processed. We find

$$u := (-1) - \lfloor 14/11 \rfloor \, 4 = -5$$

$$t := 4.$$

Therefore, to the user, *gcdext* returns the values $g = 1, u = -5, t = 4$, and we see that the procedure has found the representation (4.3.1) in this case. The importance of the 'trivial case' where $m = 0$ is apparent.

Exercises for section 4.3

1. Give a complete formal proof of theorem 4.3.1. Your proof should be by induction (on what?) and should use the extended Euclidean algorithm.

2. Find integers t, u such that

    (a) $1 = 4t + 7u$

    (b) $1 = 24t + 35u$

    (c) $5 = 65t + 100u$

3. Let $a_1, \ldots, a_n$ be positive integers.

    (a) How would you compute $gcd(a_1, \ldots, a_n)$?

    (b) Prove that there exist integers $t_1, \ldots, t_n$ such that

$$gcd(a_1, \ldots, a_n) = t_1 a_1 + t_2 a_2 + \cdots + t_n a_n.$$

    (c) Give a recursive algorithm for the computation of $t_1, \ldots, t_n$ in part (b) above.

4. If $r = ta + ub$, where r, a, b, u, v are all integers, must $r = gcd(a, b)$? What, if anything, can be said about the relationship of r to $gcd(a, b)$?

5. Let (t_0, u_0) be one pair of integers t, u for which $gcd(a, b) = ta + ub$. Find *all* such pairs of integers, a and b being given.

6. Find *all* solutions to exercises 2(a)-(c) above.

7. Find the multiplicative inverse of 49 modulo 73, using the extended Euclidean algorithm.

8. If *gcdext* is called with $(n, m) = (98, 30)$, draw a picture of the complete tree of calls that will occur during the recursive execution of the program. In your picture show, for each recursive call in the tree, the values of the input parameters to that call and the values of the output variables that were returned by that call.

4.4 Primality testing

In Chapter 1 we discussed the important distinction between algorithms that run in polynomial time *vs.* those that may require exponential time. Since then we have seen some fast algorithms and some slow ones. In the network flow problem the complexity of the MPM algorithm was $O(V^3)$, a low power of the size of the input data string, and the same holds true for the various matching and connectivity problems that are special cases of the network flow algorithm.

Likewise, the Fast Fourier Transform is really Fast. It needs only $O(n \log n)$ time to find the transform of a sequence of length n if n is a power of two, and only $O(n^2)$ time in the worst case, where n is prime.

In both of those problems we were dealing with computational situations near the low end of the complexity scale. It is feasible to do a Fast Fourier Transform on, say, 1000 data points. It is feasible to calculate maximum flows in networks with 1000 vertices or so.

On the other hand, the recursive computation of the chromatic polynomial in section 2.3 of Chapter 2 was an example of an algorithm that might use exponential amounts of time.

In this chapter we will meet another computational question for which, to date, no one has ever been able to provide a polynomial-time algorithm, nor has anyone been able to prove that such an algorithm does not exist.

The problem is just this: *Given a positive integer n. Is n prime?*

The reader should now review the discussion in Example 3 of section 0.2. In that example we showed that the obvious methods of testing for primality are slow in the sense of complexity theory. That is, we do an amount of work that is an exponentially growing function of the length of the input bit string if we use one of those methods. So this problem, which seems like a 'pushover' at first glance, turns out to be extremely difficult.

Although it is not known if a polynomial-time primality testing algorithm exists, remarkable progress on the problem has been made in recent years.

One of the most important of these advances was made independently and almost simultaneously by Solovay and Strassen, and by Rabin, in 1976-7. These authors took the imaginative step of replacing 'certainly' by 'probably,' and they devised what should be called a probabilistic compositeness (an integer is *composite* if it is *not* prime) test for integers, that runs in polynomial time.

Here is how the test works. First choose a number b uniformly at random, $1 \le b \le n - 1$. Next, subject the pair (b, n) to a certain test, called a *pseudoprimality test*, to be described below. The test has two possible outcomes: either the number n is correctly declared to be composite or the test is inconclusive.

If that were the whole story it would be scarcely have been worth the telling. Indeed the test 'Does b divide n?' already would perform the function stated above. However, it has a low probability of success even if n is composite, and if the answer is 'No,' we would have learned virtually nothing.

The additional property that the test described below has, not shared by the more naive test 'Does b divide n?,' is that *if n is composite, the chance that the test will declare that result is at least 1/2.*

In practice, for a given n we would apply the test 100 times using 100 numbers b_i that are independently chosen at random in $[1, n - 1]$. If n is composite, the probability that it will be declared composite at least once is at least $1 - 2^{-100}$, and these are rather good odds. Each test would be done in quick polynomial time. If n is not found to be composite after 100 trials, and if certainty is important, then it would be worthwhile to subject n to one of the nonprobabilistic primality tests in order to dispel all doubt.

It remains to describe the test to which the pair (b, n) is subjected, and to prove that it detects compositeness with probability $\ge 1/2$.

Before doing this we mention another important development. A more recent primality test, due to Adleman, Pomerance and Rumely in 1983, is completely deterministic. That is, given n it will *surely* decide whether or not n is prime. The test is more elaborate than the one that we are about to describe, and it runs in tantalizingly close to polynomial time. In fact it was shown to run in time

$$O\big((\log n)^{c \log \log \log n}\big)$$

for a certain constant c. Since the number of bits of n is a constant multiple of $\log n$, this latter estimate is of the form

$$O\big((Bits)^{c \log \log Bits}\big).$$

The exponent of '$Bits$,' which would be constant in a polynomial time algorithm, in fact grows extremely slowly as n grows. This is what was referred to as 'tantalizingly close' to polynomial time, earlier.

It is important to notice that in order to prove that a number is not prime, it is certainly *sufficient* to find a nontrivial divisor of that number. It is not *necessary* to do that, however. All we are asking for is a 'yes' or 'no' answer to the question 'is n prime?.' If you should find it discouraging to get only the answer 'no' to the question 'Is 7122643698294074179 prime?,' without getting any of the factors of that number, then what you want is a fast algorithm for the factorization problem.

In the test that follows, the decision about the compositeness of n will be reached without a knowledge of any of the factors of n. This is true of the Adleman, Pomerance, Rumely test also. The question of finding a factor of n, or all of them, is another interesting computational problem that is under active investigation. Of course the factorization problem is at least as hard as finding out if an integer is prime, and so no polynomial-time algorithm is known for it either. Again, there are probabilistic algorithms for the factorization problem just as there are for primality testing, but in the case of the factorization problem, even they don't run in polynomial-time.

In section 4.9 we will discuss a probabilistic algorithm for factoring large integers, after some motivation in section 4.8, where we remark on the connection between computationally intractable problems and cryptography. Specifically, we will describe one of the 'Public Key' data encryption systems whose usefulness stems directly from the difficulty of factoring large integers.

Isn't it amazing that in this technologically enlightened age we still don't know how to find a divisor of a whole number quickly?

4.5 Interlude: the ring of integers modulo n

In this section we will look at the arithmetic structure of the integers modulo some fixed integer n. These results will be needed in the sequel, but they are also of interest in themselves and have numerous applications.

Consider the ring whose elements are $0, 1, 2, \ldots, n-1$ and in which we do addition, subtraction, and multiplication modulo n. This ring is called $\mathbf{Z}_n$. For example, in Table 4.5.1 we show the addition and multiplication tables of $\mathbf{Z}_6$.

+	0	1	2	3	4	5		*	0	1	2	3	4	5
0	0	1	2	3	4	5		0	0	0	0	0	0	0
1	1	2	3	4	5	0		1	0	1	2	3	4	5
2	2	3	4	5	0	1		2	0	2	4	0	2	4
3	3	4	5	0	1	2		3	0	3	0	3	0	3
4	4	5	0	1	2	3		4	0	4	2	0	4	2
5	5	0	1	2	3	4		5	0	5	4	3	2	1

Table 4.5.1: Arithmetic in the ring $\mathbf{Z}_6$

Notice that while $\mathbf{Z}_n$ is a ring, it certainly need not be a *field*, because there will usually be some noninvertible elements. Reference to Table 4.5.1 shows that 2, 3, 4 have no multiplicative inverses in $\mathbf{Z}_6$, while 1, 5 do have such inverses. The difference, of course, stems from the fact that 1 and 5 are relatively prime to the modulus 6 while 2, 3, 4 are not. We learned, in corollary 4.3.1, that an element m of $\mathbf{Z}_n$ is invertible if and only if m and n are relatively prime.

The invertible elements of $\mathbf{Z}_n$ form a multiplicative *group*. We will call that group the *group of units* of $\mathbf{Z}_n$ and will denote it by U_n. It has exactly $\phi(n)$ elements, by lemma 4.5.1, where ϕ is the Euler function of (4.1.5).

The multiplication table of the group U_{18} is shown in Table 4.5.2.

Notice that U_{18} contains $\phi(18) = 6$ elements, that each of them has an inverse and that each row (column) of the multiplication table contains a permutation of all of the group elements.

Let's look at the table a little more closely, with a view to finding out

*	1	5	7	11	13	17
1	1	5	7	11	13	17
5	5	7	17	1	11	13
7	7	17	13	5	1	11
11	11	1	5	13	17	7
13	13	11	1	17	7	5
17	17	13	11	7	5	1

Table 4.5.2: Multiplication modulo 18

if the group U_{18} is *cyclic*. In a cyclic group there is an element a whose powers $1, a, a^2, a^3, \ldots$ run through all of the elements of the group.

If we refer to the table again, we see that in U_{18} the powers of 5 are $1, 5, 7, 17, 13, 11, 1, \ldots$. Thus the *order* of the group element 5 is equal to the order of the group, and the powers of 5 exhaust all group elements. The group U_{18} is indeed cyclic, and 5 is a generator of U_{18}.

A number (like 5 in the example) whose powers run through all elements of U_n is called a *primitive root* modulo n. Thus 5 is a primitive root modulo 18. The reader should now find, from Table 4.5.2, *all* of the primitive roots modulo 18.

Alternatively, since the order of a group element must always divide the order of the group, every element of U_n has an order that divides $\phi(n)$. The primitive roots are exactly the elements, if they exist, of maximum possible order $\phi(n)$.

We pause to note two corollaries of these remarks, namely

Theorem 4.5.1 ('Fermat's theorem'). *For every integer b that is relatively prime to n we have*

$$b^{\phi(n)} \equiv 1 \pmod{n}. \qquad (4.5.1)$$

In particular, if n is a prime number then $\phi(n) = n - 1$, and we have

Theorem 4.5.2 ('Fermat's little theorem'). *If n is prime, then for all $b \not\equiv 0 \pmod{n}$ we have $b^{n-1} \equiv 1 \pmod{n}$.*

It is important to know which groups U_n are cyclic, *i.e.*, which integers n have primitive roots. The answer is given by

Theorem 4.5.3. *An integer n has a primitive root if and only if $n = 2$ or $n = 4$ or $n = p^a$ (p an odd prime) or $n = 2p^a$ (p an odd prime). Hence, the groups U_n are cyclic for precisely such values of n.* ■

The proof of theorem 4.5.3 is a little lengthy and is omitted. It can be found, for example, in the book of LeVeque that is cited at the end of this chapter.

According to theorem 4.5.3, for example, U_{18} is cyclic, which we have already seen, and U_{12} is not cyclic, which the reader should check.

Further, we state as an immediate consequence of theorem 4.5.3,

Corollary 4.5.3. *If n is an odd prime, then U_n is cyclic, and in particular the equation $x^2 = 1$, in U_n, has only the solutions $x = \pm 1$.*

Next we will discuss the fact that if the integer n can be factored in the form $n = p_1^{a_1} p_2^{a_2} \cdots p_r^{a_r}$ then the full ring $\mathbf{Z}_n$ can also be factored, in a certain sense, as a 'product' of $Z_{p_i^{a_i}}$.

Let's take $\mathbf{Z}_6$ as an example. Since $6 = 2 \cdot 3$, we expect that somehow $\mathbf{Z}_6 = \mathbf{Z}_2 \otimes \mathbf{Z}_3$. What this means is that we consider *ordered pairs* x_1, x_2, where $x_1 \in \mathbf{Z}_2$ and $x_2 \in \mathbf{Z}_3$.

Here is how we do the arithmetic with the ordered pairs.

First, $(x_1, x_2) + (y_1, y_2) = (x_1 + y_1, x_2 + y_2)$, in which the two '+' signs on the right are different: the first '$x_1 + y_1$' is done in $\mathbf{Z}_2$ while the '$x_2 + y_2$' is done in $\mathbf{Z}_3$.

Second, $(x_1, x_2) \cdot (y_1, y_2) = (x_1 \cdot y_1, x_2 \cdot y_2)$, in which the two multiplications on the right side are different: the '$x_1 \cdot y_1$' is done in $\mathbf{Z}_2$ and the '$x_2 \cdot y_2$' in $\mathbf{Z}_3$.

Therefore the 6 elements of $\mathbf{Z}_6$ are

$$(0,0), (0,1), (0,2), (1,0), (1,1), (1,2).$$

A sample of the addition process is

$$(0,2) + (1,1) = (0+1, 2+1)$$
$$= (1,0)$$

where the addition of the first components was done modulo 2 and of the second components was done modulo 3.

A sample of the multiplication process is

$$(1,2) \cdot (1,2) = (1 \cdot 1, 2 \cdot 2)$$
$$= (1,1)$$

in which multiplication of the first components was done modulo 2 and of the second components was done modulo 3.

In full generality we can state the factorization of $\mathbf{Z}_n$ as

Theorem 4.5.4. *Let $n = p_1^{a_1} p_2^{a_2} \cdots p_r^{a_r}$. The mapping which associates with each $x \in \mathbf{Z}_n$ the r-tuple $(x_1, x_2, \ldots, x_r)$, where $x_i = x \bmod p_i^{a_i}$ ($i = 1, r$), is a ring isomorphism of $\mathbf{Z}_n$ with the ring of r-tuples $(x_1, x_2, \ldots, x_r)$ in which*

(a) $x_i \in \mathbf{Z}_{p_i^{a_i}}$ ($i = 1, r$) and

(b) $(x_1, \ldots, x_r) + (y_1, \ldots, y_r) = (x_1 + y_1, \ldots, x_r + y_r)$ and

(c) $(x_1, \ldots, x_r) \cdot (y_1, \ldots, y_r) = (x_1 \cdot y_1, \ldots, x_r \cdot y_r)$

(d) In (b), the ith '+' sign on the right side is the addition operation of $\mathbf{Z}_{p_i^{a_i}}$ and in (c) the ith '·' sign is the multiplication operation of $\mathbf{Z}_{p_i^{a_i}}$, for each $i = 1, 2, \ldots, r$.

The proof of theorem 4.5.4 follows at once from the famous

Theorem 4.5.5 ('The Chinese Remainder Theorem'). *Let m_i ($i = 1, r$) be pairwise relatively prime positive integers, and let*

$$M = m_1 m_2 \cdots m_r.$$

Then the mapping that associates with each integer x ($0 \leq x \leq M - 1$) the r-tuple $(b_1, b_2, \ldots, b_r)$, where $b_i = x \bmod m_i$ ($i = 1, r$), is a bijection between $\mathbf{Z}_M$ and $\mathbf{Z}_{m_1} \times \cdots \times \mathbf{Z}_{m_r}$.

A good theorem deserves a good proof. An outstanding theorem deserves two proofs, at least, one existential, and one constructive. So here are one of each for the Chinese Remainder Theorem.

Proof 1: We must show that each r-tuple $(b_1, \ldots, b_r)$ such that $0 \leq b_i < m_i$ ($i = 1, r$) occurs exactly once. There are obviously M such vectors, and so it will be sufficient to show that each of them occurs *at most* once as the image of some x.

In the contrary case we would have x and x' both corresponding to $(b_1, b_2, \ldots, b_r)$, say. But then $x - x' \equiv 0$ modulo each of the m_i. Hence $x - x'$ is divisible by $M = m_1 m_2 \cdots m_r$. But $|x - x'| < M$, hence $x = x'$. ∎

Proof 2: Here's how to compute a number x that satisfies the simultaneous congruences $x \equiv b_i \bmod m_i (i = 1, r)$. First, by the extended Euclidean algorithm we can quickly find $t_1, \ldots, t_r, u_1, \ldots, u_r$, such that $t_j(M/m_j) + u_j m_j = 1$ for $j = 1, \ldots, r$. Then we claim that the number $x = \sum_j b_j t_j(M/m_j)$ satisfies all of the given congruences. Indeed, for each $k = 1, 2, \ldots, r$ we have

$$
\begin{aligned}
x &= \sum_{j=1}^{r} b_j t_j(M/m_j) \\
&\equiv b_k t_k(M/m_k) \pmod{m_k} \\
&\equiv b_k \pmod{m_k}
\end{aligned}
$$

where the first congruence holds because each $M/m_j \, (j \neq k)$ is divisible by m_k, and the second congruence follows since

$$
t_k(M/m_k) = 1 - u_k m_k \equiv 1 \bmod m_k,
$$

completing the second proof of the Chinese Remainder Theorem. ∎

Now the proof of theorem 4.5.4 follows easily, and is left as an exercise for the reader.

The factorization that is described in detail in theorem 4.5.4 will be written symbolically as

$$
\mathbf{Z}_n \cong \bigotimes_{i=1}^{r} \mathbf{Z}_{p_i^{a_i}}. \tag{4.5.2}
$$

The factorization (4.5.2) of the ring $\mathbf{Z}_n$ induces a factorization

$$
U_n \cong \bigotimes_{i=1}^{r} U_{p_i^{a_i}} \tag{4.5.3}
$$

of the group of units. Since U_n is a group, (4.5.3) is an isomorphism of the multiplicative structure only. In $\mathbf{Z}_{12}$, for example, we find

$$
U_{12} \cong U_4 U_3
$$

155

where $U_4 = \{1,3\}$, $U_3 = \{1,2\}$. So U_{12} can be thought of as the set $\{(1,1,),(1,2),(3,1),(3,2)\}$, together with the componentwise multiplication operation described above.

Exercises for section 4.5

1. Give a complete proof of theorem 4.5.4.
2. Find all primitive roots modulo 18.
3. Find all primitive roots modulo 27.
4. Write out the multiplication table of the group U_{27}.
5. Which elements of $\mathbf{Z}_{11}$ are squares?
6. Which elements of $\mathbf{Z}_{13}$ are squares?
7. Find all $x \in U_{27}$ such that $x^2 = 1$. Find all $x \in U_{15}$ such that $x^2 = 1$.
8. Prove that if there is a primitive root modulo n then the equation $x^2 = 1$ in the group U_n has only the solutions $x = \pm 1$.
9. Find a number x that is congruent to 1, 7 and 11 to the respective moduli 5, 11 and 17. Use the method in the second proof of the remainder theorem 4.5.5.
10. Write out the complete proof of the 'immediate' corollary 4.5.3.

4.6 Pseudoprimality tests

In this section we will discuss various tests that might be used for testing the compositeness of integers probabilistically.

By a *pseudoprimality test* we mean a test that is applied to a pair (b,n) of integers, and that has the following characteristics:
(a) The possible outcomes of the test are 'n is composite' or 'inconclusive.'
(b) If the test reports 'n is composite' then n is composite.
(c) The test runs in a time that is polynomial in $\log n$.

If the test result is 'inconclusive' then we say that *n is pseudoprime to the base b* (which means that n is so far acting like a prime number, as far as we can tell).

The outcome of the test of the primality of n depends on the base b that is chosen. In a good pseudoprimality test there will be many bases b that will give the correct answer. More precisely, a good pseudoprimality test will, with high probability (*i.e.*, for a large number of choices of the base b) declare that a composite n is composite. In more detail, we will say that a pseudoprimality test is 'good' if there is a fixed positive number t such that every composite integer n is declared to be composite for at least tn choices of the base b, in the interval $1 \leq b \leq n$.

Of course, given an integer n, it is silly to say that 'there is a high probability that n is prime.' Either n is prime or it isn't, and we should not blame our ignorance on n itself. Nonetheless, the abuse of language is sufficiently appealing that we will define the problem away: we will say that a given integer n is *very probably prime* if we have subjected it to a good pseudoprimality test, with a large number of different bases b, and have found that it is pseudoprime to all of those bases.

Here are four examples of pseudoprimality tests, only one of which is 'good.'

Test 1. *Given b, n. Output 'n is composite' if b divides n, else 'inconclusive.'*

This isn't the good one. If n is composite, the probability that it will be so declared is the probability that we happen to have found a b that divides n, where b is not 1 or n. The probability of this event, if b is chosen uniformly at random from $[1, n]$, is

$$p_1 = (d(n) - 2)/n$$

where $d(n)$ is the number of divisors of n. Certainly p_1 is not bounded from below by a positive constant t, if n is composite.

Test 2. *Given b, n. Output 'n is composite' if $gcd(b, n) \neq 1$, else output 'inconclusive.'*

This one is a little better, but not yet good. If n is composite, the number of bases $b \leq n$ for which Test 2 will produce the result 'composite' is $n - \phi(n)$, where ϕ is the Euler totient function, of (4.1.5). This number of useful bases will be large if n has some small prime factors, but in that case it's easy to find out that n is composite by other methods. If n has only a few large prime factors, say if $n = p^2$, then the proportion of useful bases is very small, and we have the same kind of inefficiency as in Test 1 above.

Now we can state the third pseudoprimality test.

Test 3. *Given b, n. (If b and n are not relatively prime or) if $b^{n-1} \not\equiv 1$ (mod n) then output 'n is composite,' else output 'inconclusive.'*

Regrettably, the test is still not 'good,' but it's a lot better than its predecessors. To cite an extreme case of its un-goodness, there exist composite numbers n, called *Carmichael numbers*, with the property that the pair (b, n) produces the output 'inconclusive' for *every* integer b in $[1, n-1]$ that is relatively prime to n. An example of such a number is $n = 1729$, which is composite $(1729 = 7 \cdot 13 \cdot 19)$, but for which Test 3 gives the result 'inconclusive' on every integer $b < 1729$ that is relatively prime to 1729 (*i.e.*, that is not divisible by 7 or 13 or 19).

Despite such misbehavior, the test usually seems to perform quite well. When $n = 169$ (a difficult integer for tests 1 and 2) it turns out that there are 158 different b's in $[1,168]$ that produce the 'composite' outcome from Test 3, namely every such b except for 19, 22, 23, 70, 80, 89, 99, 146, 147, 150, 168.

Finally, we will describe a good pseudoprimality test. The familial resemblance to Test 3 will be apparent.

Test 4. *(the strong pseudoprimality test): Given (b, n). Let $n - 1 = 2^q m$, where m is an odd integer. If either*
(a) $b^m \equiv 1 \pmod{n}$ or
(b) there is an integer i in $[0, q-1]$ such that

$$b^{m2^i} \equiv -1 \pmod{n}$$

then return 'inconclusive' else return 'n is composite.'

First we validate the test by proving the

Proposition. *If the test returns the message 'n is composite,' then n is composite.*

Proof: Suppose not. Then n is an odd prime. We claim that

$$b^{m2^i} \equiv 1 \pmod{n}$$

for all $i = q, q-1, \ldots, 0$. If so then the case $i = 0$ will contradict the outcome of the test, and thereby complete the proof. To establish the claim, it is clearly true when $i = q$, by Fermat's theorem. If true for i, then it is true for $i - 1$ also, because

$$(b^{m2^{i-1}})^2 = b^{m2^i}$$

$$\equiv 1 \pmod{n}$$

implies that the quantity being squared is $+1$ or -1. Since n is an odd prime, by corollary 4.5.3 U_n is cyclic, and so the equation $x^2 = 1$ in U_n has only the solutions $x = \pm 1$. But -1 is ruled out by the outcome of the test, and the proof of the claim is complete. ■

What is the computational complexity of the test? Consider first the computational problem of raising a number to a power. We can calculate, for example, $b^m \bmod n$ with $O(\log m)$ integer multiplications, by successive squaring. More precisely, we compute b, b^2, b^4, $b^8, \ldots$ by squaring, and reducing modulo n immediately after each squaring operation, rather than waiting until the final exponent is reached. Then we use the binary expansion of the exponent m to tell us which of these powers of b we should multiply together in order to compute b^m. For instance,

$$b^{337} = b^{256} \cdot b^{64} \cdot b^{16} \cdot b.$$

The complete power algorithm is recursive and looks like this:

```
function power(b, m, n);
{returns bᵐ mod n}
if m = 0
   then
      power := 1
   else
      t := sqr(power(b, ⌊m/2⌋, n));
      if m is odd  then t := t · b;
      power := t mod n
end.{power}
```

Hence part (a) of the strong pseudoprimality test can be done in $O(\log m) = O(\log n)$ multiplications of integers of at most $O(\log n)$ bits each. Similarly, in part (b) of the test there are $O(\log n)$ possible values of i to check, and for each of them we do a single multiplication of two integers each of which has $O(\log n)$ bits (this argument, of course, applies to Test 3 above also).

The entire test requires, therefore, some low power of $\log n$ bit operations. For instance, if we were to use the most obvious way to multiply two B bit numbers we would do $O(B^2)$ bit operations, and then the above test would take $O((\log n)^3)$ time. This is a polynomial in the number of bits of input.

In the next section we are going to prove that Test 4 is a good pseudoprimality test in that if n is composite then at least half of the integers b, $1 \le b \le n - 1$ will give the result 'n is composite.'

For example, if $n = 169$, then it turns out that for 157 of the possible 168 bases b in $[1,168]$, Test 4 will reply '169 is composite.' The only bases b that 169 can fool are 19, 22, 23, 70, 80, 89, 99, 146, 147, 150, 168. For this case of $n = 169$ the performances of Test 4 and of Test 3 are identical. However, there are no analogues of the Carmichael numbers for Test 4.

Exercises for section 4.6

1. Given an odd integer n. Let $T(n)$ be the set of all $b \in [1, n]$ such that $gcd(b, n) = 1$ and $b^{n-1} \equiv 1 \pmod{n}$. Show that $|T(n)|$ divides $\phi(n)$.

2. Let H be a cyclic group of order n. How many elements of each order r are there in H (r divides n)?

3. If $n = p^a$, where p is an odd prime, then the number of $x \in U_n$ such that x has exact order r, is $\phi(r)$, for all divisors r of $\phi(n)$. In particular, the number of primitive roots modulo n is $\phi(\phi(n))$.

4. If $n = p_1^{a_1} \cdots p_m^{a_m}$, and if r divides $\phi(n)$, then the number of $x \in U_n$ such that $x^r \equiv 1 \pmod{n}$ is

$$\prod_{i=1}^{m} gcd(\phi(p_i^{a_i}), r).$$

5. In a group G suppose f_m and g_m are, respectively, the number of elements of order m and the number of solutions of the equation $x^m = 1$, for each $m = 1, 2, \ldots$. What is the relationship between these two sequences? That is, how would you compute the g's from the f's? the f's from the g's? If you have never seen a question of this kind, look in any book on the theory of numbers, find 'Möbius inversion,' and apply it to this problem.

4.7 Proof of goodness of the strong pseudoprimality test

In this section we will show that if n is composite, then at least half of the integers b in $[1, n-1]$ will yield the result 'n is composite' in the strong pseudoprimality test. The basic idea of the proof is that a subgroup of a group that is not the entire group can consist of at most half of the elements of that group.

Suppose n has the factorization

$$n = p_1^{a_1} \cdots p_s^{a_s}$$

and let $n_i = p_i^{a_i}$ $(i = 1, s)$.

Lemma 4.7.1. *The order of each element of U_n is a divisor of $e^* = lcm\{\phi(n_i);\ i = 1, s\}$.*

Proof: From the product representation (4.5.3) of U_n we find that an element x of U_n can be regarded as an s-tuple of elements from the cyclic groups U_{n_i} $(i = 1, s)$. The order of x is equal to the lcm of the orders of the elements of the s-tuple. But for each $i = 1, \ldots, s$ the order of the ith of those elements is a divisor of $\phi(n_i)$, and therefore the order of x divides the *lcm* shown above. ■

Lemma 4.7.2. *Let $n > 1$ be odd. For each element u of U_n let $C(u) = \{1, u, u^2, \ldots, u^{e-1}\}$ denote the cyclic group that u generates. Let B be the set of all elements u of U_n for which $C(u)$ either contains -1 or has odd order (e odd). If B generates the full group U_n then n is a prime power.*

Proof: Let $e^* = 2^t m$, where m is odd and e^* is as shown in lemma 4.7.1. Then there is a j such that $\phi(n_j)$ is divisible by 2^t.

Now if n is a prime power, we are finished. So we can suppose that n is divisible by more than one prime number. Since $\phi(n)$ is an even number for all $n > 2$ (proof?), the number e^* is even. Hence $t > 0$ and we can define a mapping ψ of the group U_n to itself by

$$\psi(x) = x^{2^{t-1} m} \qquad (x \in U_n)$$

(note that $\psi(x)$ is its own inverse).

This is in fact a group homomorphism:

$$\forall x, y \in U_n : \ \psi(xy) = \psi(x)\psi(y).$$

Let B be as in the statement of lemma 4.7.2. For each $x \in B$, $\psi(x)$ is in $C(x)$ and

$$\psi(x)^2 = \psi(x^2) = 1.$$

Since $\psi(x)$ is an element of $C(x)$ whose square is 1, $\psi(x)$ has order 1 or 2. Hence if $\psi(x) \neq 1$, it is of order 2. If the cyclic group $C(x)$ is of odd order then it contains no element of even order. Hence $C(x)$ is of even order and contains -1. Then it can contain no other element of order 2, so $\psi(x) = -1$ in this case.

Hence *for every $x \in B$, $\psi(x) = \pm 1$.*

Suppose B generates the full group U_n. Then not only for every $x \in B$ but *for every $x \in U_n$ it is true that $\psi(x) = \pm 1$.*

Suppose n is not a prime power. Then $s > 1$ in the factorization (4.5.2) of U_n. Consider the element v of U_n which, when written out as an s-tuple according to that factorization, is of the form

$$v = (1, 1, 1, \ldots, 1, y, 1, \ldots, 1)$$

where the 'y' is in the jth component, $y \in U_{n_j}$ (recall that j is as described above, in the second sentence of this proof). We can suppose y to be an element of order exactly 2^t in U_{n_j} since U_{n_j} is cyclic.

Consider $\psi(v)$. Clearly $\psi(v)$ is not 1, for otherwise the order of y, namely 2^t, would divide $2^{t-1}m$, which is impossible because m is odd.

Also, $\psi(v)$ is not -1, because the element -1 of U_n is represented uniquely by the s-tuple all of whose entries are -1. Thus $\psi(v)$ is neither 1 nor -1 in U_n, which contradicts the italicized assertion above. Hence $s = 1$ and n is a prime power, completing the proof. ■

Now we can prove the main result of Solovay, Strassen and Rabin, which asserts that Test 4 is good.

Theorem 4.7.1. *Let B' be the set of integers $b \bmod n$ such that (b, n) returns 'inconclusive' in Test 4.*

(a) If B' generates U_n then n is prime.

(b) If n is composite then B′ consists of at most half of the integers in [1, n − 1].

Proof: Suppose $b \in B'$ and let m be the odd part of $n - 1$. Then either $b^m \equiv 1$ or $b^{m2^i} \equiv -1$ for some $i \in [0, q - 1]$. In the former case the cyclic subgroup $C(b)$ has odd order, since m is odd, and in the latter case $C(b)$ contains -1.

Hence in either case $B' \subseteq B$, where B is the set defined in the statement of lemma 4.7.2 above. If B' generates the full group U_n then B does too, and by lemma 4.7.2, n is a prime power, say $n = p^k$.

Also, in either of the above cases we have $b^{n-1} \equiv 1$, so the same holds for all $b \in B'$, and so for all $x \in U_n$ we have $x^{n-1} \equiv 1$, since B' generates U_n.

Now U_n is cyclic of order

$$\phi(n) = \phi(p^k) = p^{k-1}(p - 1).$$

By theorem 4.5.3 there are primitive roots modulo $n = p^k$. Let g be one of these. The order of g is, on the one hand, $p^{k-1}(p - 1)$ since the set of all of its powers is identical with U_n, and on the other hand is a divisor of $n - 1 = p^k - 1$ since $x^{n-1} \equiv 1$ for all x, and in particular for $x = g$.

Hence $p^{k-1}(p - 1)$ (which, if $k > 1$, is a multiple of p) divides $p^k - 1$ (which is one less than a multiple of p), and so $k = 1$, which completes the proof of part (a) of the theorem.

In part (b), n is composite and so B' cannot generate all of U_n, by part (a). Hence B' generates a proper subgroup of U_n, and so can contain at most half as many elements as U_n contains, and the proof is complete. ∎

Another application of the same circle of ideas to computer science occurs in the generation of random numbers on a computer. A good way to do this is to choose a primitive root modulo the word size of your computer, and then, each time the user asks for a random number, output the next higher power of the primitive root. The fact that you started with a primitive root insures that the number of 'random numbers' generated before repetition sets in will be as large as possible.

Now we'll summarize the way in which the primality test is used. Suppose there is given a large integer n, and we would like to determine if it is prime.

We would do

> function $testn(n, outcome)$;
> $times := 0$;
> **repeat**
> > choose an integer b uniformly at random in $[2, n-1]$;
> > apply the strong pseudoprimality test (Test 4) to the
> > > pair (b, n);
> > $times := times + 1$
> **until** {result is 'n is composite' or $times = 100$};
> **if** $times = 100$ **then** $outcome$:='n probably prime'
> > > **else** $outcome$:='n is composite'
> end{$testn$}

If the procedure exits with 'n is composite,' then we can be certain that n is not prime. If we want to see the factors of n then it will be necessary to use some factorization algorithm, such as the one described below in section 4.9.

On the other hand, if the procedure halts because it has been through 100 trials without a conclusive result, then the integer n is very probably prime. More precisely, the chance that a composite integer n would have behaved like that is less than 2^{-100}. If we want certainty, however, it will be necessary to apply a test whose outcome will *prove* primality, such as the algorithm of Adleman, Rumely and Pomerance, referred to earlier.

In section 4.9 we will discuss a probabilistic factoring algorithm. Before doing so, in the next section we will present a remarkable application of the complexity of the factoring problem, to cryptography. Such applications remind us that primality and factorization algorithms have important applications beyond pure mathematics, in areas of vital public concern.

Exercises for section 4.7

1. For $n = 9$ and for $n = 15$ find all of the cyclic groups $C(u)$, of lemma 4.7.2, and find the set B.

2. For $n = 9$ and $n = 15$ find the set B', of theorem 4.7.1.

4.8 Factoring and cryptography

A computationally intractable problem can be used to create secure codes for the transmission of information over public channels of communication. The idea is that those who send the messages to each other will have extra pieces of information that will allow them to solve the intractable problem rapidly, whereas an aspiring eavesdropper would be faced with an exponential amount of computation.

Even if we don't have a provably computationally intractable problem, we can still take a chance that those who might intercept our messages won't know any polynomial-time algorithms if we don't know any. Since there are precious few *provably* hard problems, and hordes of *apparently* hard problems, it is scarcely surprising that a number of sophisticated coding schemes rest on the latter rather than the former. One should remember, though, that an adversary might discover fast algorithms for doing these problems and keep that fact secret while deciphering all of our messages.

A remarkable feature of a family of recently developed coding schemes, called 'Public Key Encryption Systems,' is that the 'key' to the code lies in the public domain, so it can be easily available to sender and receiver (and eavesdropper), and can be readily changed if need be. On the negative side, the most widely used Public Key Systems lean on computational problems that are only *presumed* to be intractable, like factoring large integers, rather than having been *proved* so.

We are going to discuss a Public Key System called the RSA scheme, after its inventors: Rivest, Shamir and Adleman. This particular method depends for its success on the seeming intractability of the problem of finding the factors of large integers. If that problem could be done in polynomial time, then the RSA system could be 'cracked.'

In this system there are three centers of information: the sender of the message, the receiver of the message, and the Public Domain (for instance, the 'Personals' ads of the *New York Times*). Here is how the system works.

(A) Who knows what and when

Here are the items of information that are involved, and who knows each item:

p, q: two large prime numbers, chosen by the receiver, and told to nobody else (not even to the sender!).

n : the product *pq* is *n*, and this is placed in the Public Domain.

E : a random integer, placed in the Public Domain by the receiver, who has first made sure that *E* is relatively prime to $(p-1)(q-1)$ by computing the g.c.d., and choosing a new *E* at random until the g.c.d. is 1. This is easy for the receiver to do because *p* and *q* are known to him, and the g.c.d. calculation is fast.

P : a message that the sender would like to send, thought of as a string of bits whose value, when regarded as a binary number, lies in the range $[0, n-1]$.

In addition to the above, one more item of information is computed by the receiver, and that is the integer *D* that is the multiplicative inverse mod $(p-1)(q-1)$ of *E*, i.e.,

$$DE \equiv 1 \pmod{(p-1)(q-1)}.$$

Again, since *p* and *q* are known, this is a fast calculation for the receiver, as we shall see.

To summarize,

> The receiver knows *p, q, D*
> The sender knows *P*
> Everybody knows *n* and *E*

In Fig. 4.8.1 we show the interiors of the heads of the sender and receiver, as well as the contents of the Public Domain.

(B) How to send a message

The sender takes the message *P*, looks at the public keys *E* and *n*, computes $C \equiv P^E \pmod{n}$, and transmits *C* over the public airwaves.

Note that the sender has no private codebook or anything secret other than the message itself.

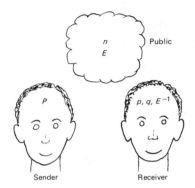

Fig. 4.8.1: Who knows what

(C) How to decode a message

The receiver receives C, and computes $C^D \mod n$. Observe, however, that $(p-1)(q-1)$ is $\phi(n)$, and so we have

$$
\begin{aligned}
C^D &\equiv P^{DE} \\
&= P^{(1+t\phi(n))} \qquad (t \text{ is some integer}) \\
&\equiv P \pmod n
\end{aligned}
$$

where the last equality is by Fermat's theorem (4.5.1). The receiver has now recovered the original message P.

If the receiver suspects that the code has been broken, *i.e.*, that the adversaries have discovered the primes p and q, then the sender can change them without having to send any secret messages to anyone else. Only the public numbers n and E would change. The sender would not need to be informed of any other changes.

Before proceeding, the reader is urged to construct a little scenario. Make up a short (very short!) message. Choose values for the other parameters that are needed to complete the picture. Send the message as the sender would, and decode it as the receiver would. Then try to intercept the message, as an eavesdropper would, and see what the difficulties are.

(D) How to intercept the message

An eavesdropper who receives the message C would be unable to decode it without (inventing some entirely new decoding scheme or) knowing the inverse D of E (mod $(p-1)(q-1)$). The eavesdropper, however, does not even know the modulus $(p-1)(q-1)$ because p and q are unknown (only the receiver knows them), and knowing the product $pq = n$ alone is insufficient. The eavesdropper is thereby compelled to derive a polynomial-time factoring algorithm for large integers. May success attend those efforts!

The reader might well remark here that the receiver has a substantial computational problem in creating two large primes p and q. To a certain extent this is so, but two factors make the task a good deal easier. First, p and q will need to have only half as many bits as n has, so the job is of smaller size. Second, there are methods that will produce large prime numbers very rapidly as long as one is not too particular about which primes they are, as long as they are large enough. We will not discuss those methods here.

The elegance of the RSA cryptosystem prompts a few more remarks that are intended to reinforce the distinction between exponential- and polynomial-time complexities.

How hard is it to factor a large integer? At this writing, integers of up to perhaps a couple of hundred digits can be approached with some confidence that factorization will be accomplished within a few hours of the computing time of a very fast machine. If we think in terms of a message that is about the length of one typewritten page, then that message would contain about 8000 bits, equivalent to about 2400 decimal digits. This is in contrast to the largest feasible length that can be handled by contemporary factoring algorithms of about 200 decimal digits. A one-page message is therefore well into the zone of computational intractability.

How hard is it to find the multiplicative inverse, $\bmod(p-1)(q-1)$? If p and q are *known* then it's easy to find the inverse, as we saw in corollary 4.3.1. Finding an inverse mod n is no harder than carrying out the extended Euclidean algorithm, *i.e.*, it's a linear time job.

4.9 Factoring large integers

The problem of finding divisors of large integers is in a much more primitive condition than is primality testing. For example, we don't even know a *probabilistic* algorithm that will return a factor of a large composite integer, with probability $> 1/2$, in polynomial time.

In this section we will discuss a probabilistic factoring algorithm that finds factors in an *average* time that is only moderately exponential, and that's about the state of the art at present.

Let n be an integer whose factorization is desired.

Definition. *By a factor base B we will mean a set of distinct nonzero integers $\{b_0, b_1, \ldots, b_h\}$.*

Definition. *Let B be a factor base. An integer a will be called a B-number if the integer c that is defined by the conditions*

> *(a) $c \equiv a^2 \pmod{n}$ and*
> *(b) $-n/2 \leq c < n/2$*

can be written as a product of factors from the factor base B.

If we let $e(a, i)$ denote the exponent of b_i in that product, then we have

$$a^2 \equiv \prod_{i=0}^{h} b_i^{e(a,i)} \pmod{n}.$$

Hence, for each B-number we get an $(h+1)$-vector of exponents $\mathbf{e}(a)$.

Suppose we can find enough B-numbers so that the resulting collection of exponent vectors is a linearly dependent set, *mod 2*. For instance, a set of $h + 2$ B-numbers would certainly have that property.

Then we could nontrivially represent the zero vector as a sum of a certain set A of exponent vectors, say

$$\sum_{a \in A} \mathbf{e}(a) \equiv (0, 0, \ldots, 0) \pmod{2}.$$

Now define the integers

$$r_i = (1/2) \sum_{a \in A} e(a, i) \quad (i = 0, 1, \ldots h)$$

$$u = \prod_A a \pmod{n}$$

$$v = \prod_i b_i^{r_i}.$$

It then would follow, after an easy calculation, that $u^2 \equiv v^2 \pmod{n}$. Hence either $u - v$ or $u + v$ has a factor in common with n. It may be, of course, that $u \equiv \pm v$ (mod n), in which case we would have learned nothing. However if neither $u \equiv v$ (mod n) nor $u \equiv -v$ (mod n) is true then we will have found a nontrivial factor of n, namely $gcd(u - v, n)$ or $gcd(u + v, n)$.

Example:

Take as a factor base $B = \{-2, 5\}$, and let it be required to find a factor of $n = 1729$. Then we claim that 186 and 267 are B-numbers. To see that 186 is a B-number, note that $186^2 = 20 \cdot 1729 + (-2)^4$, and similarly, since $267^2 = 41 \cdot 1729 + (-2)^4 5^2$, we see that 267 is a B-number, for this factor base B.

The exponent vectors of 186 and 167 are $(4, 0)$ and $(4, 2)$ respectively, and these sum to $(0, 0)$ (mod 2), hence we find that

$$u = 186 \times 267 \equiv 1250 \quad (\text{mod } 1729)$$

$$r_1 = 4; \quad r_2 = 1$$

$$v = (-2)^4 (5)^1 = 80$$

$$gcd(u - v, n) = gcd(1170, 1729) = 13$$

and we have found the factor 13 of 1729. ■

There might have seemed to be some legerdemain involved in plucking the B-numbers 186 and 267 out of the air, in the example above. In fact, as the algorithm has been implemented by its author, J. D. Dixon, one simply chooses integers uniformly at random from $[1, n-1]$ until enough B-numbers have been found so their exponent vectors are linearly dependent modulo 2. In Dixon's implementation the factor base that is used consists of -1 together with the first h prime numbers.

It can then be proved that if n is not a prime power then with a correct choice of h relative to n, if we repeat the random choices until a factor of n is found, the average running time will be

$$exp\{(2 + o(1))(\log \log \log n)^{.5}\}.$$

This is not polynomial time, but it is moderately exponential only. Nevertheless, it is close to being about the best that we know how to do on the elusive problem of factoring a large integer.

4.10 Proving primality

In this section we will consider a problem that sounds a lot like primality testing, but is really a little different because the rules of the game are different. Basically the problem is to convince a skeptical audience that a certain integer is prime, requiring them to do only a small amount of computation in order to be so persuaded.

First, though, suppose you were writing a 100-decimal-digit integer n on the blackboard in front of a large audience and you wanted to prove to them that n was *not* a prime.

If you simply wrote down two smaller integers whose product was n, the job would be done. Anyone who wished to be certain could spend a few minutes multiplying the factors together and verifying that their product was indeed n, and all doubts would be dispelled.

Indeed*, a speaker at a mathematical convention in 1903 announced the result that $2^{67} - 1$ is not a prime number, and to be utterly convincing all he had to do was to write

$$2^{67} - 1 = 193707721 \times 761838257287.$$

We note that the speaker probably had to work very hard to *find* those factors, but having found them it became quite easy to convince others of the truth of the claimed result.

A pair of integers r, s for which $r \neq 1$, $s \neq 1$, and $n = rs$ constitute a *certificate* attesting to the compositeness of n. With this certificate $C(n)$ and an auxiliary checking algorithm, *viz.*

(1) Verify that $r \neq 1$, and that $s \neq 1$

(2) Verify that $rs = n$

we can prove, in polynomial time, that n is not a prime number.

Now comes the hard part. How might we convince an audience that a certain integer n *is* a prime number? The rules are that we are allowed to do any immense amount of calculation beforehand, and the results of that calculation can be written on a certificate $C(n)$ that accompanies the integer n. The audience, however, will need to do only a polynomial amount of further computation in order to convince themselves that n is prime.

* We follow the account given in V. Pratt, Every prime has a succinct certificate, *SIAM J. Computing*, 4 (1975), 214-220.

We will describe a primality-checking algorithm $\mathcal{A}$ with the following properties:

(1) Inputs to $\mathcal{A}$ are the integer n and a certain certificate $C(n)$.

(2) If n is prime then the action of $\mathcal{A}$ on the inputs $(n, C(n))$ results in the output 'n is prime.'

(3) If n is not prime then *for every possible certificate* $C(n)$ the action of $\mathcal{A}$ on the inputs $(n, C(n))$ results in the output 'primality of n is not verified.'

(4) Algorithm $\mathcal{A}$ runs in polynomial time.

Now the question is, does such a procedure exist for primality verification? The answer is affirmative, and we will now describe one. The fact that primality can be quickly verified, if not quickly discovered, is of great importance for the developments of Chapter 5. In the language of section 5.1, what we are about to do is to show that the problem 'Is n prime?' belongs to the class NP.

The next lemma is a kind of converse to 'Fermat's little theorem' (theorem 4.5.2).

Lemma 4.10.1. *Let p be a positive integer. Suppose there is an integer x such that $x^{p-1} \equiv 1 \pmod{p}$ and such that for all divisors d of $p-1$, $d < p-1$, we have $x^d \not\equiv 1 \pmod{p}$. Then p is prime.*

Proof: First we claim that $gcd(x, p) = 1$, for let $g = gcd(x, p)$. Then $x = gg'$, $p = gg''$. Since $x^{p-1} \equiv 1 \pmod{p}$ we have $x^{p-1} = 1 + tp$ and $x^{p-1} - tp = (gg')^{p-1} - tgg'' = 1$. The left side is a multiple of g. The right side is not, unless $g = 1$.

It follows that $x \in U_p$, the group of units of $\mathbf{Z}_p$. Thus x is an element of order $p-1$ in a group of order $\phi(p)$. Hence $(p-1)|\phi(p)$. But always $\phi(p) \le p-1$. Hence $\phi(p) = p-1$ and p is prime. ∎

Lemma 4.10.1 is the basis for V. Pratt's method of constructing certificates of primality. The construction of the certificate is actually *recursive* since step 3^0 below calls for certificates of smaller primes. We suppose that the certificate of the prime 2 is the trivial case, and that it can be verified at no cost.

Here is a complete list of the information that is on the certificate $C(p)$ that accompanies an integer p whose primality is to be attested to:

1^0: a list of the primes p_i and the exponents a_i for the canonical factorization $p - 1 = \prod_{i=1}^{r} p_i^{a_i}$

2^0: the certificates $C(p_i)$ of each of the primes $p_1, \ldots, p_r$

3^0: a positive integer x.

To verify that p is prime we could execute the following algorithm $\mathcal{B}$:

(B1) Check that $p - 1 = \prod p_i^{a_i}$.

(B2) Check that each p_i is prime, using the certificates $C(p_i)$ $(i = 1, r)$.

(B3) For each divisor d of $p - 1$, $d < p - 1$, check that $x^d \not\equiv 1$ (mod p).

(B4) Check that $x^{p-1} \equiv 1$ (mod p).

This algorithm $\mathcal{B}$ is correct, but it might not operate in polynomial time. In step B3 we are looking at every divisor of $p - 1$, and there may be a lot of them.

Fortunately, it isn't necessary to check *every* divisor of $p-1$. The reader will have no trouble proving that *there is a divisor d of $p - 1$ $(d < p - 1)$ for which $x^d \equiv 1$ (mod p) if and only if there is such a divisor that has the special form $d = (p - 1)/p_i$.*

The primality checking algorithm $\mathcal{A}$ now reads as follows.

(A1) Check that $p - 1 = \prod p_i^{a_i}$.

(A2) Check that each p_i is prime, using the certificates $C(p_i)$ $(i = 1, r)$.

(A3) For each $i := 1$ to r, check that

$$x^{(p-1)/p_i} \not\equiv 1 \quad (\text{mod } p).$$

(A4) Check that $x^{p-1} \equiv 1$ (mod p).

Now let's look at the complexity of algorithm $\mathcal{A}$.

We will measure its complexity by the number of times that we have to do a computation of either of the types (a) 'is $m = \prod q_j^{b_j}$?' or (b) 'is $y^s \equiv 1$ (mod p)?'

Let $f(p)$ be that number. Then we have (remembering that the algorithm calls itself r times)

$$f(p) = 1 + \sum_{i=2}^{r} f(p_i) + r + 1 \qquad (4.10.1)$$

in which the four terms, as written, correspond to the four steps in the checking algorithm. The sum begins with '$i = 2$' because the prime 2, which is a lways a divisor of $p - 1$, is 'free.'

Now (4.10.1) can be written as

$$g(p) = \sum_{i=2}^{r} g(p_i) + 4 \qquad (4.10.2)$$

where $g(p) = 1 + f(p)$. We claim that $g(p) \le 4\log_2 p$ for all p.

This is surely true if $p = 2$. If true for primes less than p then from (4.10.2),

$$g(p) \le \sum_{i=2}^{r} \{4\log_2 p_i\} + 4$$

$$= 4\log_2 \{\prod_{i=2}^{r} p_i\} + 4$$

$$\le 4\log_2 \{(p-1)/2\} + 4$$

$$= 4\log_2 (p-1)$$

$$\le 4\log_2 p.$$

Hence $f(p) \le 4\log_2 p - 1$ for all $p \ge 2$. ■

Since the number of bits in p is $\Theta(\log p)$, the number $f(p)$ is a number of executions of steps that is a polynomial in the length of the input bit string. We leave to the exercises the verification that each of the steps that $f(p)$ counts is also executed in polynomial time, so the entire primality-verification procedure operates in polynomial time. This yields

Theorem 4.10.1. *(V. Pratt, 1975) There exist a checking algorithm and a certificate such that primality can be verified in polynomial time.*

Exercises for section 4.10

1. Show that two positive integers of b bits each can be multiplied with at most $O(b^2)$ bit operations (multiplications and carries).

2. Prove that step A1 of algorithm $\mathcal{A}$ can be executed in polynomial time, where time is now measured by the number of bit operations that are implied by the integer multiplications.

3. Same as exercise 2 above, for steps A3 and A4.

4. Write out the complete certificate that attests to the primality of 19.

5. Find an upper bound for the total number of bits that are in the certificate of the integer p.

6. Carry out the complete checking algorithm on the certificate that you prepared in exercise 4 above.

7. Let $p = 15$. Show that there is no integer x as described in the hypotheses of lemma 4.10.1.

8. Let $p = 17$. Find all integers x that satisfy the hypotheses of lemma 4.10.1.

Bibliography

The material in this chapter has made extensive use of the excellent review article

John D. Dixon, Factorization and primality tests, *The American Mathematical Monthly*, **91** (1984), 333-352.

A basic reference for number theory, Fermat's theorem, etc. is

G. H. Hardy and E. M. Wright, *An Introduction to the Theory of Numbers*, Oxford University Press, Oxford, 1954.

Another is

W. J. LeVeque, *Fundamentals of Number Theory*, Addison-Wesley, Reading, MA, 1977.

The probabilistic algorithm for compositeness testing was found by

M. O. Rabin, Probabilistic algorithms, in *Algorithms and Complexity, New Directions and Recent Results*, J. Traub ed., Academic Press, New York, 1976.

and at about the same time by

R. Solovay and V. Strassen, A fast Monte Carlo test for primality, *SIAM Journal of Computing*, **6** (1977), pp. 84-85; *erratum ibid.*, **7** (1978), 118.

Some empirical properties of that algorithm are in

C. Pomerance, J. L. Selfridge and S. Wagstaff Jr., The pseudoprimes to $25 \cdot 10^9$, *Mathematics of Computation*, **35** (1980), 1003-1026.

The fastest nonprobabilistic primality test appeared first in

L. M. Adleman, On distinguishing prime numbers from composite numbers, *IEEE Abstracts*, May 1980, 387-406.

A more complete account, together with the complexity analysis, is in

L. M. Adleman, C. Pomerance and R. S. Rumely, On distinguishing prime numbers from composite numbers, *Annals of Mathematics* **117** (1983), 173-206.

A streamlined version of the above algorithm was given by

H. Cohen and H. W. Lenstra Jr., Primality testing and Jacobi sums, Report 82-18, Math. Inst. U. of Amsterdam, Amsterdam, 1982.

The idea of public key data encryption is due to

W. Diffie and M. E. Hellman, New directions in cryptography, *IEEE Transactions on Information Theory*, **IT-22**, 6 (1976), 644-654.

An account of the subject is contained in

M. E. Hellman, The mathematics of public key cryptography, *Scientific American*, **241**, 2 (August 1979), 146-157.

The use of factoring as the key to the code is due to

R. L. Rivest, A. Shamir and L. M. Adleman, A method for obtaining digital signatures and public key cryptosystems, *Communications of the A.C.M.*, **21**, 2 (February 1978), 120-126.

The probabilistic factoring algorithm in the text is that of

John D. Dixon, Asymptotically fast factorization of integers, *Mathematics of Computation*, **36** (1981), 255-260.

Chapter 5: NP-completeness

5.1 Introduction

In the previous chapter we met two computational problems for which fast algorithms have never been found, but neither have such algorithms been proved to be unattainable. Those were the primality-testing problem, for which the best-known algorithm is delicately poised on the brink of polynomial time, and the integer-factoring problem, for which the known algorithms are in a more primitive condition.

In this chapter we will meet a large family of such problems (hundreds of them now!). This family is not just a list of seemingly difficult computational problems. It is in fact bound together by strong structural ties. The collection of problems, called the *NP-complete* problems, includes many well-known and important questions in discrete mathematics, such as the following.

The travelling salesman problem ('TSP'): Given n points in the plane ('cities'), and a distance D. Is there a tour that visits all n of the cities, returns to its starting point, and has total length $\leq D$?

Graph coloring: Given a graph G and an integer K. Can the vertices of G be properly colored in K or fewer colors?

Independent set: Given a graph G and an integer K. Does $V(G)$ contain an independent set of K vertices?

Bin packing: Given a finite set S of positive integers, an integer K (the 'bin capacity') and an integer N (the number of bins). Does there exist a partition of S into N or fewer subsets such that the sum of the integers in each subset is $\leq K$? In other words, can we 'pack' the integers of S into at most N 'bins,' where the 'capacity' of each bin is K?

These are very difficult computational problems. Take the graph coloring problem, for instance. We could try every possible way of coloring the vertices of G in K colors to see if any of them work. There are K^n such possibilities if G has n vertices. Hence a very large amount of computation will be done, enough so that if G has 50 vertices and we have 10 colors at our disposal, the problem would lie far beyond the capabilities of the fastest computers that are now available.

Hard problems can have easy instances. If the graph G happens to have no edges at all, or very few of them, then it will be very easy to find out if a coloring is possible, or if an independent set of K vertices is present.

The real question is this (let's use 'Independent Set' as an illustration). Is it possible to design an algorithm that will come packaged with a performance guarantee of the following kind:

> **The seller warrants that if a graph G, of n vertices, and a positive integer K are input to this program, then it will correctly determine if there is an independent set of K or more vertices in $V(G)$, and it will do so in an amount of time that is at most $1000n^8$ minutes.**

Hence there is no contradiction between the facts that the problem is hard and that there are easy cases. The hardness of the problem stems from the seeming impossibility of producing such an algorithm accompanied by such a manufacturer's warranty card. Of course the '$1000n^8$' didn't have to be exactly that. But some quite specific polynomial in the length of the input bit string must appear in the performance guarantee. Hence '$357n^9$' might have appeared in the guarantee, and so might '$23n^3$,' but 'n^K' would not be allowed.

Let's look carefully at why n^K would not be an acceptable worst-case polynomial time performance bound. In the 'Independent Set' problem the input must describe the graph G and the integer K. How many bits are needed to do that? The graph can be specified, for example, by its vertex adjacency matrix A. This is an $n \times n$ matrix in which the entry in row i and column j is 1 if $(i, j) \in E(G)$ and is 0 else.

Evidently n^2 bits of input will describe the matrix A. The integers K and n can be entered with just $O(\log n)$ bits, so the entire input bit string for the 'Independent Set' problem is $\sim n^2$ bits long. Let B denote the number of bits in the input string. Suppose that on the warranty card the program was guaranteed to run in a time T that is $< n^K$.

Is this a guarantee of polynomial-time performance? That question means 'Is there a polynomial P such that for every instance of 'Independent Set' the running time T will be at most $P(B)$?' Well, is T bounded by a polynomial in B if $T = n^K$ and $B \sim n^2$? It would seem so; in fact obviously $T = O(B^{K/2})$, and that's a polynomial, isn't it?

The key point resides in the order of the qualifiers. We must give the polynomial that works for *every* instance of the problem *first*. Then that one single polynomial must in fact work on every instance. If the 'polynomial' that we give is $B^{K/2}$, well that's a different polynomial in B for different instances of the problem, because K is different for different instances. Therefore if we say that a certain program for 'Independent Set' will always get an answer before $B^{K/2}$ minutes, where B is the length of the input bit string, then we would not have provided a polynomial-time guarantee in the required form of a single polynomial in B that applies uniformly to all problem instances.

The distinction is a little thorny, but is worthy of careful study because it's of fundamental importance. What we are discussing is usually called a *worst-case* time bound, meaning a bound on the running time that applies to every instance of the problem. Worst-case time bounds aren't the only possible interesting ones. Sometimes we might not care if an algorithm is occasionally very slow as long as it is *almost always fast*. In other situations we might be satisfied with an algorithm that is fast *on the average*. For the present, however, we will stick to the worst-case time bounds and study some of the theory that applies to that situation. In sections 5.6 and 5.7 we will study some average time bounds.

Now let's return to the properties of the NP-complete family of problems. Here are some of them.

1^0: The problems all seem to be computationally very difficult, and no polynomial time algorithms have been found for any of them.

2^0: It has not been proved that polynomial time algorithms for these problems do not exist.

3^0: But this is not just a random list of hard problems. *If a fast algorithm could be found for one NP-complete problem then there would be fast algorithms for all of them.*

4^0: Conversely, if it could be proved that no fast algorithm exists for one of the NP-complete problems, then there could not be a fast algorithm for any other of those problems.

The above properties are not intended to be a *definition* of the concept of NP-completeness. We'll get to that later on in this section. They are intended as a list of some of the interesting features of these problems which, when coupled with their theoretical and practical importance, accounts for the intense worldwide research effort that has gone into understanding them in recent years.

The question of the existence or nonexistence of polynomial-time algorithms for the NP-complete problems probably rates as the principal unsolved problem that faces theoretical computer science today.

Our next task will be to develop the formal machinery that will permit us to give precise definitions of all of the concepts that are needed. In the remainder of this section we will discuss the additional ideas informally, and then in section 5.2 we'll state them quite precisely.

What is a decision problem?

First, the idea of a *decision problem*. A decision problem is one that asks only for a yes-or-no answer: Can this graph be 5-colored? Is there a tour of length < 15 miles? Is there a set of 67 independent vertices?

Many of the problems that we are studying can be phrased as decision problems or as *optimization problems*: What is the smallest number of colors with which G can be colored? What is the length of the shortest tour of these cities? What is the size of the largest independent set of vertices in G?

Usually, if we find a fast algorithm for a decision problem then with just a little more work we will be able to solve the corresponding optimization problem. For instance, suppose we have an algorithm that solves the decision problem for graph coloring, and what we want is the solution of the optimization problem (the chromatic number).

Let a graph G be given, of, say, 100 vertices. Ask: can the graph be 50-colored? If so, then the chromatic number lies between 1 and 50. Then ask if it can be colored in 25 colors. If not, then the chromatic number lies between 26 and 50. Continue in this way, using bisection of the interval that is known to contain the chromatic number. After $O(\log n)$ steps we will have found the chromatic number of a graph of n vertices. The extra multiplicative factor of $\log n$ will not alter the polynomial vs. nonpolynomial running time distinction. Hence, if there is a fast way to do the decision problem then there is a fast way to do the optimization problem. The converse is obvious.

Hence we will restrict our discussion to decision problems.

What is a language?

Since every decision problem can have only the two answers 'Y/N,' we can think of a decision problem as asking if a given word (the input string) does or does not belong to a certain *language*. The language is the totality of words for which the answer is 'Y.'

The graph 3-coloring language, for instance, is the set of all symmetric, square matrices of 0,1 entries, with zeros on the main diagonal (these are the vertex adjacency matrices of graphs) such that the graph that the matrix represents is 3-colorable. We can imagine that there is somewhere a vast dictionary of all of the words in this language. A 3-colorablity computation is therefore nothing but an attempt to discover whether a given word belongs to the dictionary.

What is the class P?

We say that a decision problem belongs to the class P if there is an algorithm A and a number c such that for every instance I of the problem the algorithm A will produce a solution in time $O(B^c)$, where B is the number of bits in the input string that represents I.

To put it more briefly, P is the set of easy decision problems.

Examples of problems in P are most of the ones that we have already met in this book: Are these two integers relatively prime? Is this integer divisible by that one? Is this graph 2-colorable? Is there a flow of value greater than K in this network? Can this graph be disconnected by the removal of K or fewer edges? Is there a matching of more than K edges in this bipartite graph? For each of these problems there is a fast (polynomial-time) algorithm.

What is the class NP?

The class NP is a little more subtle. A decision problem Q belongs to NP if there is an algorithm A that does the following:

(a) Associated with each word of the language Q (*i.e.*, with each instance of I for which the answer is 'Yes') there is a *certificate* $C(I)$ such that when the pair $(I, C(I))$ are input to algorithm A it recognizes that I belongs to the language Q.

(b) If I is some word that does not belong to the language Q then there is no choice of certificate $C(I)$ that will cause $\mathcal{A}$ to recognize I as a member of the language Q.

(c) Algorithm $\mathcal{A}$ operates in polynomial time.

To put this one more briefly, NP is the class of decision problems for which it is easy to *check* the correctness of a claimed answer, with the aid of a little extra information. So we aren't asking for a way to *find* a solution, but only to *verify* that an alleged solution really is correct.

Here is an analogy that may help to clarify the distinction between the classes P and NP. We have all had the experience of reading through a truly ingenious and difficult proof of some mathematical theorem, and wondering how the person who found the proof in the first place ever did it. Our task, as a reader, was only to *verify* the proof, and that is a much easier job than the mathematician who invented the proof had. To pursue the analogy just a bit farther, some proofs are extremely time consuming even to check (see the proof of the four-color theorem!), and similarly, some computational problems are not even known to belong to NP, let alone to P.

In P are the problems where it's easy to *find* a solution, and in NP are the problems where it's easy to *check* a solution that may have been very tedious to find.

Here's another example. Consider the graph-coloring problem to be the decision problem Q. Certainly this problem is not known to be in P. It is, however, in NP, and here is an algorithm and a method of constructing certificates that proves it.

Suppose G is some graph that *is* K-colorable. The certificate of G might be a list of the colors that get assigned to each vertex of G in some proper K-coloring of the vertices of G. Where did we get that list, you ask? Well, we never said it was easy to construct a certificate. If you actually want to find one, then you will really have to solve a hard problem. But we're really only talking about *checking* the correctness of an alleged answer. To *check* that a certain graph G really is K-colorable we can be convinced if you will show us the color of each vertex in a proper K-coloring.

If you do provide that certificate, then our checking algorithm $\mathcal{A}$ is very simple. It checks first that every vertex has a color and only one color. It then checks that no more than K colors have been used altogether. It finally checks that for each edge e of G it is true that the two endpoints of e have different colors.

Hence the graph-coloring problem belongs to NP.

For the travelling salesman problem we would provide a certificate that contains a tour, whose total length is $\leq K$, of all of the cities. The checking algorithm $\mathcal{A}$ would then verify that the tour really does visit all of the cities and really does have total length $\leq K$.

The travelling salesman problem, therefore, also belongs to NP.

'Well,' you might reply, 'if we're allowed to look at the answers, how could a problem fail to belong to NP?'

Try this decision problem: an instance I of the problem consists of a set of n cities in the plane and a positive number K. The question is 'Is it true that there is *not* a tour of all of these cities whose total length is less than K?' Clearly this is a kind of negation of the travelling salesman problem. Does it belong to NP? If so, there must be an algorithm $\mathcal{A}$ and a way of making a certificate $C(I)$ for each instance I such that we can quickly verify that *no such tour exists* of the given cities. Any suggestions for the certificate? The algorithm? No one else knows how to do this either.

It is not known if this negation of the travelling salesman problem belongs to NP.

Are there problems that *do* belong to NP but for which it isn't immediately obvious that this is so? Yes. In fact that's one of the main reasons that we studied the algorithm of Pratt, in section 4.10. Pratt's algorithm is exactly a method of producing a certificate with the aid of which we can quickly check that a given integer is prime. The decision problem 'Given n, is it prime?' is thereby revealed to belong to NP, although that fact wasn't obvious at a glance.

It is very clear that P$\subseteq$NP. Indeed, if $Q \in$P is some decision problem then we can verify membership in the language Q with the empty certificate. That is, we don't even need a certificate in order to do a quick calculation that checks membership in the language because the problem itself can be quickly solved.

It seems natural to suppose that NP is larger than P. That is, one might presume that there are problems whose solutions can be quickly checked with the aid of a certificate even though they can't be quickly found in the first place.

No example of such a problem has ever been produced (and proved), nor has it been proved that no such problem exists. The question of whether or not P=NP is the one that we cited earlier as being perhaps the most important open question in the subject area today.

It is fairly obvious that the class P is called 'the class P' because 'P' is the first letter of 'Polynomial Time.' But what does 'NP' stand for? Stay tuned; the answer will appear in section 5.2.

What is reducibility?

Suppose that we want to solve a system of 100 simultaneous linear equations in 100 unknowns, of the form $Ax = b$. We run down to the local software emporium and quickly purchase a program for $49.95 that solves such systems. When we get home and read the fine print on the label we discover, to our chagrin, that the program works only on systems where the matrix A is symmetric, and the coefficient matrix in the system that we want to solve is, of course, not symmetric.

One possible response to this predicament would be to look for the solution to the system $A^T Ax = A^T b$, in which the coefficient matrix $A^T A$ is now symmetric.

What we would have done would be to have *reduced* the problem that we are really interested in to an instance of a problem for which we have an algorithm.

More generally, let Q and Q' be two decision problems. We will say that Q' *is quickly reducible to* Q if whenever we are given an instance I' of the problem Q' we can convert it, with only a polynomial amount of labor, into an instance I of Q, in such a way that I' and I both have the same answer ('Yes' or 'No').

Thus, if we buy a program to solve Q, then we can use it to solve Q', with just a small amount of extra work.

What is NP-completeness?

How would you like to buy one program, for $49.95, that can solve 500 different kinds of problems? That's what NP-completeness is about.

To state it a little more carefully, *a decision problem is NP-complete if it belongs to NP and every problem in NP is quickly reducible to it.*

The implications of NP-completeness are numerous. Suppose we could prove that a certain decision problem Q is NP-complete. Then we could concentrate our efforts to find polynomial-time algorithms on just that one problem Q. Indeed, if we were to succeed in finding a polynomial-time algorithm to do instances of Q then we would automatically have found a fast algorithm for doing every problem in NP. How does that work?

Take an instance I' of any problem Q' in NP. Since Q' is quickly reducible to Q we could transform the instance I' into an instance I of Q. Then use the super algorithm that we found for problems in Q to decide I. Altogether only a polynomial amount of time will have been used from start to finish.

Let's be more specific. Suppose that tomorrow morning we prove that the graph coloring problem is NP-complete, and that on the next morning you find a fast algorithm for solving it. Then, consider some instance of the bin-packing problem. Since graph-coloring is NP-complete, the instance of bin-packing can be quickly converted into an instance of graph-coloring for which the 'Yes/No' answer is the same. Now use the fast graph-coloring algorithm that you found (congratulations, by the way!) on the converted problem. The answer you get is the correct answer for the original bin-packing problem.

So, a fast algorithm for some NP-complete problem implies a fast algorithm for every problem in NP. Conversely, suppose we can prove that it is impossible to find a fast algorithm for some particular problem Q in NP. Then we can't find a fast algorithm for any NP-complete problem Q' either. For if we could then we would be able to solve instances of Q by quickly reducing them to instances of Q' and solving them.

If we could prove that there is no fast way to test the primality of a given integer then we would have proved that there is no fast way to decide if graphs are K-colorable, because, as we will see, the graph-coloring problem is NP-complete and primality testing is in NP. Think about that one for a few moments, and the extraordinary beauty and structural unity of these computational problems will begin to reveal itself.

To summarize: quick for one NP-complete problem implies quick for all of NP; provably slow for one problem in NP implies provably slow for all NP-complete problems.

There's just one small detail to attend to. We've been discussing the economic advantages of keeping flocks of unicorns instead of sheep. If there aren't any unicorns then the discussion is a little silly.

NP-complete problems have all sorts of marvellous properties. It's lovely that every problem in NP can be quickly reduced to just that one NP-complete problem. *But are there any NP-complete problems?* Why, after all, should there be a single computational problem with the property that every one of the diverse creatures that inhabit NP should be quickly reducible to it?

Well, there *are* NP-complete problems, hordes of them, and proving that will occupy our attention for the next two sections. Here's the plan.

In section 5.2 we are going to talk about a simple computer, called a Turing machine. It is an idealized computer, and its purpose is to standardize ideas of computability and time of computation by referring all problems to the one standard machine.

A Turing machine is an extremely simple finite-state computer, and when it performs a computation, a unit of computational labor will be very clearly and unambiguously describable. It turns out that the important aspects of polynomial-time computability do not depend on the particular computer that is chosen as the model. The beauty of the Turing machine is that it is at once a strong enough concept that it can in principle perform any calculation that any other finite-state machine can do, while at the same time it is logically clean and simple enough to be useful for proving theorems about complexity.

The microcomputer on your desktop *might* have been chosen as the standard against which polynomial-time computability is measured. If that had been done then the class P of quickly solvable problems would scarcely have changed at all (the polynomials would be different, but they would still be polynomials), but the *proofs* that we humans would have to give in order to establish the relevant theorems would have gotten much more complicated because of the variety of different kinds of states that modern computers have.

Next, in section 5.3 we will prove that there *is* an NP-complete problem. It is called *the satisfiability problem.* Its status as an NP-complete problem was established by S. Cook in 1971, and from that work all later progress in this field has flowed. The proof uses the theory of Turing machines.

The first NP-complete problem was the hardest one to find. We will find, in section 5.4, a few more NP-complete problems, so the reader will get some idea of the methods that are used in identifying them.

Since nobody knows a fast way to solve these problems, various methods have been developed that give approximate solutions quickly, or that give exact solutions in fast *average* time, and so forth. The beautiful book of Garey and Johnson (see references at the end of the chapter) calls this 'coping with NP-completeness,' and we will spend the rest of this chapter discussing some of those ideas.

Exercises for section 5.1

1. Prove that the following decision problem belongs to P: Given $a_1, \ldots, a_n$ and K, all integers. Is the median of the a's smaller than K?

2. Prove that the following decision problem is in NP: given an $n \times n$ matrix of integer entries. Is det $A = 0$?

3. For which of the following problems can you prove membership in P?

    (a) Given a graph G. Does G contain a circuit of length 4?

    (b) Given a graph G. Is G bipartite?

    (c) Given n integers. Is there a subset of them whose sum is an even number?

    (d) Given n integers. Is there a subset of them whose sum is divisible by 3?

    (e) Given a graph G. Does G contain an Euler circuit?

4. For which of the following problems can you prove membership in NP?

    (a) Given a set of integers and another integer K. Is there a subset of the given integers whose sum is K?

    (b) Given a graph G and an integer K. Does G contain a path of length $\geq K$?

    (c) Given a set of K integers. Is it true that not all of them are prime?

    (d) Given a set of K integers. Is it true that all of them are prime?

5.2 Turing Machines

A Turing machine consists of

(a) a doubly infinite *tape*, that is marked off into *squares* that are numbered as shown in Fig. 5.2.1 below. Each square can contain a single character from the character set that the machine recognizes. For simplicity, we can assume that the character set contains just three symbols: '0,' '1,' and ' ' (blank).

(b) a *tape head* that is capable of either reading a single character from a square on the tape or writing a single character on a square, or moving its position relative to the tape by an increment of one square in either direction.

(c) a finite list of *states* such that at every instant the machine is in exactly one of those states. The possible states of the machine are, first of all, the regular states $q_1, q_2, \ldots, q_s$ and, second, three *special states*

q_0: the initial state

q_Y: the final state in a problem to which the answer is 'Yes'

q_N: the final state in a problem to which the answer is 'No'

(d) a *program* (or *program module*, if we think of it as a pluggable component) that directs the machine through the steps of a particular task.

Fig. 5.2.1: A Turing machine tape

Let's describe the program module in more detail. Suppose that at a certain instant the machine is in a state q (other than q_Y or q_N) and that the symbol that has just been read from the tape is '*symbol*.' Then from the pair $(q, symbol)$ the program module will decide

(i) to what state q' the machine shall next go and

(ii) what single character the machine will now write on the tape in the square over which the head is now positioned and

(iii) whether the tape head will next move one square to the right or one square to the left.

One step of the program, therefore, goes from

$$(state,\ symbol) \text{ to } (newstate,\ newsymbol,\ increment). \qquad (5.2.1)$$

If and when the state reaches q_Y or q_N the computation is over and the machine halts.

The machine should be thought of as part hardware and part software. The programmer's job is, as usual, to write the software. To write a program for a Turing machine what we have to do is to tell it how to make each and every one of the transitions (5.2.1). A Turing machine program looks like a table in which, for every possible pair (*state*, *symbol*) that the machine might find itself in, the programmer has specified what the *newstate*, the *newsymbol* and the *increment* shall be.

To begin a computation with a Turing machine we take the input string x, of length B, say, that describes the problem that we want to solve, and we write x in squares $1, 2, \ldots, B$ of the tape. The tape head is then positioned over square 1, the machine is put into state q_0, the program module that the programmer prepared is plugged into its slot, and the computation begins.

The machine reads the symbol in square 1. It now is in state q_0 and has read *symbol*, so it can consult the program module to find out what to do. The program instructs it to write at square 1 a *newsymbol*, to move the head either to square 0 or to square 2, and to enter a certain *newstate*, say q'. The whole process is then repeated, possibly forever, but hopefully after finitely many steps the machine will enter either state q_Y or state q_N, at which moment the computation will halt with the decision having been made.

If we want to watch a Turing machine in operation, we don't have to build it, we can simulate one. Here is a pidgin-Pascal simulation of a Turing machine that can easily be turned into a functioning program. It is in two principal parts.

The procedure *turmach* has for input a string x of length B, and for output it sets the Boolean variable *accept* to *True* or *False*, depending on whether the outcome of the computation is that the machine halted in state q_Y or q_N, respectively. This procedure is the 'hardware' part of the Turing machine. It doesn't vary from one job to the next.

Procedure *gonextto* is the program module of the machine, and it will be different for each task. Its inputs are the present *state* of the machine and the *symbol* that was just read from the tape. Its outputs are the *newstate* into which the machine goes next, the *newsymbol* that the tape head now writes on the current square, and the *increment* (± 1) by which the tape head will now move.

```
    procedure turmach(B:integer; x:array[1..B]; accept:Boolean);
    {simulates Turing machine action on input string x of length B}
    {write input string on tape in first B squares}
     for square := 1 to B do
         tape[square] := x[square];
    {record boundaries of written-on part of tape}
     leftmost:=1; rightmost:=B;
    {initialize tape head and state}
     state:=0; square := 1;
     while state ≠ 'Y' and state ≠ 'N' do
          {read symbol at current tape square}
          if square < leftmost or square > rightmost
              then symbol:=' '  else symbol:= tape[square];
          {ask program module for state transition}
          gonextto(state, symbol, newstate, newsymbol, increment);
          state :=newstate;
          {update boundaries and write new symbol};
            if square > rightmost  then rightmost := square;
            if square < leftmost  then leftmost := square;
            tape[square] := newsymbol;
          {move tape head}
            square := square + increment
      end; {while}
      accept := { state = 'Y' }
    end.{turmach}
```

Now let's try to write a particular program module *gonextto*. Consider the following problem: given an input string x, consisting of 0's and 1's, of length B. Find out if it is true that the string contains an odd number of 1's.

We will write a program that will scan the input string from left to right, and at each moment the machine will be in state 0 if it has so far scanned an even number of 1's, in state 1 otherwise. In Fig. 5.2.2 we show a program that will get the job done.

Exercise. Program the above as procedure *gonextto*, run it for some input

state	symbol	newstate	newsymbol	increment
0	0	0	0	+1
0	1	1	1	+1
0	*blank*	q_N	*blank*	-1
1	0	1	0	+1
1	1	0	1	+1
1	*blank*	q_Y	*blank*	-1

Fig. 5.2.2: A Turing machine program for bit parity

string, and print out the state of the machine, the contents of the tape, and the position of the tape head after each step of the computation.

In the next section we are going to use the Turing machine concept to prove Cook's theorem, which is the assertion that a certain problem is NP-complete. Right now, let's review some of the ideas that have already been introduced from the point of view of Turing machines.

We might immediately notice that some terms that were just a little fuzzy before are now much more sharply in focus. Take the notion of polynomial time, for example. To make that idea precise one needs a careful definition of what 'the length of the input bit string' means, and what one means by the number of 'steps' in a computation.

But on a Turing machine both of those ideas come through with crystal clarity. The input bit string x is what we write on the tape to get things started, and its length is the number of tape squares it occupies. A 'step' in a Turing machine calculation is obviously a single call to the program module. A Turing machine calculation was done 'in time $P(B)$' if the input string occupied B tape squares and the calculation took $P(B)$ steps.

Another word that we have been using without ever nailing down precisely is 'algorithm.' We all understand informally what an algorithm is. But now we understand formally too. An algorithm for a problem is a program module for a Turing machine that will cause the machine to halt after finitely many steps in state 'Y' for every instance whose answer is 'Yes,' and after finitely many steps in state 'N' for every instance whose answer is 'No.'

A Turing machine and an algorithm define a *language*. The language is the set of all input strings x that lead to termination in state 'Y,' *i.e.*, to an *accepting* calculation.

Now let's see how the idea of a Turing machine can clarify the description of the class NP. This is the class of problems for which the decisions can be made quickly if the input strings are accompanied by suitable certificates.

By a *certificate* we mean a finite strip of Turing machine tape, consisting of 0 or more squares, each of which contains a symbol from the character set of the machine in each square. A certificate can be loaded into a Turing machine as follows. If the certificate contains $m > 0$ tape squares, then replace the segment from square number $-m$ to square number -1 inclusive of the Turing machine tape with the certificate. The information on the certificate is then available to the program module just as any other information on the tape is available.

To use a Turing machine as a *checking* or *verifying* computer, we place the input string x that describes the problem-instance in squares $1, 2, \ldots, B$ of the tape, and we place the certificate $C(x)$ of x in squares $-m, -m + 1, \ldots, -1$ of the tape. We then write a verifying program for the program module in which the program verifies that the string x is indeed a word in the language of the machine, and in the course of the verification the program is quite free to examine the certificate as well as the problem instance.

A Turing machine that is being used as a *verifying* computer is called a *nondeterministic* machine. The hardware is the same, but the manner of input and the question that is being asked are different from the situation with a *deterministic* Turing machine, in which we decide whether or not the input string is in the language without using any certificates.

The class NP ('Nondeterministic Polynomial') consists of those decision problems for which there exists a fast (polynomial-time) algorithm that will verify, given a problem instance string x and a suitable certificate $C(x)$, that x belongs to the language recognized by the machine, and for which, if x does not belong to the language, *no* certificate would cause an accepting computation to ensue.

5.3 Cook's Theorem

The NP-complete problems are the hardest problems in NP, in the sense that if Q' is any decision problem in NP and Q is an *NP*-complete problem, then every instance of Q' is polynomially reducible to an instance of Q. As we have already remarked, the surprising thing is that there is an NP-complete problem at all, since it is not immediately clear why any single problem should hold the key to the polynomial-time solvability of every problem in the class NP. But there is one. As soon as we see why there is one, then we'll be able to see more easily why there are hundreds of them, including many famous computational questions about discrete structures such as graphs, networks, and games and about optimization problems, about algebraic structures, formal logic, and so forth.

Here is the *satisfiability problem*, the first problem that was proved to be NP-complete, by Stephen Cook in 1971.

We begin with a list of (Boolean) variables $x_1, x_2, ..., x_n$. A *literal* is either one of the variables x_i or the negation of one of the variables, as $\overline{x}_i$. There are $2n$ possible literals.

A *clause* is a set of literals.

The rules of the game are these. We assign the value '*True*' (T) or '*False*' (F), to each one of the *variables*. Having done that, each one of the *literals* inherits a truth value, namely a literal x_i has the same truth or falsity as the corresponding variable x_i, and a literal $\overline{x}_i$ has the opposite truth value from that of the variable x_i.

Finally, each of the clauses also inherits a truth value from this process, and it is determined as follows. A clause has the value 'T' if and only if *at least one* of the literals in that clause has the value 'T,' and otherwise it has the value 'F.'

Hence, starting with an assignment of truth values to the variables, some true and some false, we end up with a determination of the truth values of each of the clauses, some true and some false.

Definition. *A set of clauses is satisfiable if there exists an assignment of truth values to the variables that makes all of the clauses true.*

Think of the word 'or' as being between each of the literals in a clause and the word 'and' as being between the clauses.

The satisfiability problem (SAT). *Given a set of clauses. Does there exist a set of truth values (= T or F), one for each variable, such that every clause contains at least one literal whose value is T (i.e., such that every clause is satisfied)?*

Example: Consider the set x_1, x_2, x_3 of variables. From these we might manufacture the following list of four clauses:

$$\{x_1, \overline{x}_2\}, \quad \{x_1, x_3\}, \quad \{x_2, \overline{x}_3\}, \quad \{\overline{x}_1, x_3\}.$$

If we choose the truth values (T,T,F) for the variables, respectively, then the four clauses would acquire the truth values (T,T,T,F), and so this would not be a *satisfying* truth assignment for the set of clauses. There are only 8 possible ways to assign truth values to 3 variables, and after a little more experimentation we might find out that these clauses would in fact be satisfied if we were to make the assignments (T,T,T) (how can we recognize a set of clauses that is satisfied by assigning to every variable the value 'T'?). ■

The example already leaves one with the feeling that SAT might be a tough computational problem, because there are 2^n possible sets of truth values that we might have to explore, if we were to do an exhaustive search.

It is quite clear, however, that this problem belongs to NP. Indeed it is a decision problem. Furthermore we can easily assign a certificate to every set of clauses for which the answer to SAT is 'Yes, the clauses are satisfiable.' The certificate contains a set of truth values, one for each variable, that satisfy all of the clauses. A Turing machine that receives the set of clauses, suitably encoded, as input, along with the above certificate, would have to verify only that if truth values are assigned to the variables as shown on the certificate then indeed every clause does indeed contain at least one literal of value T. That verification is certainly a polynomial-time computation.

Now comes the hard part. We want to show

Theorem 5.3.1. *(S. Cook, 1971): SAT is NP-complete.*

Before we carry out the proof, it may be helpful to give a small example of the reducibility ideas that we are going to use.

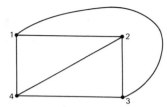

Fig. 5.3.1: A 3-coloring problem

Example. *Reducing graph-coloring to SAT.*

Consider the graph G of 4 vertices that is shown in Fig. 5.3.1, and the decision problem 'Can the vertices of G be properly colored in 3 colors?'

Let's see how that decision problem can be reduced to an instance of SAT. We will use 12 Boolean variables: the variable $x_{i,j}$ corresponds to the assertion that 'vertex i has been colored in color j' $(i = 1, 2, 3, 4; j = 1, 2, 3)$.

The instance of SAT that we construct has 31 clauses. The first 16 of these are

$$
\begin{aligned}
C(i) &:= \{x_{i,1}, x_{i,2}, x_{i,3}\} & (\forall i := 1, 2, 3, 4) \\
T(i) &:= \{\overline{x}_{i,1}, \overline{x}_{i,2}\} & (\forall i := 1, 2, 3, 4) \\
U(i) &:= \{\overline{x}_{i,1}, \overline{x}_{i,3}\} & (\forall i := 1, 2, 3, 4) \\
V(i) &:= \{\overline{x}_{i,2}, \overline{x}_{i,3}\} & (\forall i := 1, 2, 3, 4).
\end{aligned}
\tag{5.3.1}
$$

In the above, the four clauses $C(i)$ assert that each vertex has been colored in at least one color. The clauses $T(i)$ say that no vertex has both color 1 and color 2. Similarly, the clauses $U(i)$ (*resp.* $V(i)$) guarantee that no vertex has been colored 1 and 3 (*resp.* 2 and 3).

All 16 of the clauses in (5.3.1) together amount to the statement that 'each vertex has been colored in one and only one of the three available colors.'

Next we have to construct the clauses that will assure us that the two endpoints of an edge of the graph are never of the same color. For this purpose we define, for each edge e of the graph G and color j (=1,2,3), a clause $D(e, j)$ as follows. Let u and v be the two endpoints of e; then $D(e, j) := \{\overline{x}_{u,j}, \overline{x}_{v,j}\}$, which asserts that not both endpoints of the edge e have the same color j.

The original instance of the graph-coloring problem has now been reduced to an instance of SAT. In more detail, *there exists an assignment of values T,F to the 12 Boolean variables $x_{1,1}, \ldots, x_{4,3}$ such that each of the above 31 clauses contains at least one literal whose value is T if and only*

196

if the vertices of the graph G can be properly colored in three colors. The graph is 3-colorable if and only if the clauses are satisfiable. ∎

It is clear that if we have an algorithm that will solve SAT, then we can also solve graph-coloring problems. A few moments' thought will convince the reader that the transformation of one problem to the other that was carried out above involves only a polynomial amount of computation, despite the seemingly large number of variables and clauses. Hence graph-coloring is polynomially reducible to SAT.

Proof of Cook's theorem

We want to prove that SAT is NP-complete, *i.e.*, that every problem in NP is polynomially reducible to an instance of SAT. Hence let Q be some problem in NP and let I be an instance of problem Q. Since Q is in NP there exists a Turing machine that recognizes encoded instances of problem Q, if accompanied by a suitable certificate, in polynomial time.

Let TMQ be such a Turing machine, and let $P(n)$ be a polynomial in its argument n with the property that TMQ recognizes every pair $(x, C(x))$, where x is a word in the language Q and $C(x)$ is its certificate, in time $\leq P(n)$, where n is the length of x.

We intend to construct, corresponding to each word I in the language Q, an instance $f(I)$ of SAT for which the answer is 'Yes, the clauses are all simultaneously satisfiable.' Conversely, if the word I is not in the language Q, the clauses will not be satisfiable.

The idea can be summarized like this: *the instance of SAT that will be constructed will be a collection of clauses that together express the fact that there exists a certificate that causes Turing machine TMQ to do an accepting calculation.* Therefore, in order to test whether or not the word Q belongs to the language, it suffices to check that the collection of clauses is satisfiable.

To construct an instance of SAT means that we are going to define a number of variables, of literals and of clauses, in such a way that the clauses are satisfiable if and only if x is in the language Q, *i.e.*, the machine TMQ accepts x and its certificate.

What we must do, then, is to express the accepting computation of the Turing machine as the simultaneous satisfaction of a number of logical propositions. It is precisely here that the relative simplicity of a Turing machine allows us to enumerate all of the possible paths to an accepting computation in a way that would be quite unthinkable with a 'real' computer.

Now we will describe the Boolean variables that will be used in the clauses under construction.

Variable $Q_{i,k}$ is true if after step i of the checking calculation it is true that the Turing machine TMQ is in state q_k, false otherwise.

Variable $S_{i,j,k} = \{$after step i, symbol k is in tape square $j\}$.

Variable $T_{i,j} = \{$after step i, the tape head is positioned over square $j\}$.

Let's count the variables that we've just introduced. Since the Turing machine TMQ does its accepting calculation in time $\leq P(n)$ it follows that the tape head will never venture more than $\pm P(n)$ squares away from its starting position. Therefore the subscript j, which runs through the various tape squares that are scanned during the computation, can assume only $O(P(n))$ different values.

The index i runs over steps of the accepting computation, and so it also takes at most $O(P(n))$ different values.

Finally, k indexes a state of the Turing machine, and there is only some fixed finite number, K, say, of states that TMQ might be in. Hence there are altogether $O(P(n)^2)$ variables, a polynomial number of them.

Is it true that every random assignment of *True* or *False* values to each of these variables corresponds to an accepting computation on $(x, C(x))$? Certainly not. For example, if we aren't careful we might assign *True* values to $T_{9,4}$ and to $T_{10,33}$, thereby burning out the bearings on the tape transport mechanism! (why?)

Our remaining task, then, will be to describe precisely the conditions under which a set of values assigned to the variables listed above actually defines a possible accepting computation for $(x, C(x))$. Then we will be sure that whatever set of satisfying values of the variables might be found by solving the SAT problem, they will determine a real accepting calculation of the machine TMQ.

This will be done by requiring that a number of clauses be all true ('satisfied') at once, where each clause will express one necessary condition. In the following, the boldface type will describe, in words, the condition that we want to express, and it will be followed by the formal set of clauses that actually expresses the condition on input to SAT.

At each step, the machine is in at least one state

Hence, at least one of the K available state variables must be *True*. This leads to the first set of clauses, one for each step i of the computation:

$$\{Q_{i,1}, ..., Q_{i,K}\}$$

Since i takes $O(P(n))$ values, these are $O(P(n))$ clauses.

At each step, the machine is not in more than one state.

Therefore, for each step i, and each pair j', j'' of distinct states, the clause

$$\{\overline{Q}_{i,j'}, \overline{Q}_{i,j''}\}$$

must be true. These are $O(P(n))$ additional clauses to add to the list, but still more are needed.

At each step, each tape square contains exactly one symbol from the alphabet of the machine.

This leads to two lists of clauses, which require, first, that there is at least one symbol in each square at each step, and second, that there are not two symbols in each square at each step. The clauses that do this are

$$\{S_{i,j,1}, S_{i,j,2}, ..., S_{i,j,r}\}$$

where r is the number of letters in the alphabet, and

$$\{\overline{S}_{i,j,k'}, \overline{S}_{i,j,k''}\}$$

for each step i, square j and pair k', k'' of distinct symbols in the alphabet of the machine.

The reader will by now have gotten the idea of how to construct the clauses, so for the next three categories we will simply list the functions that must be performed by the corresponding lists of clauses, and leave the construction of the clauses as an exercise.

At each step, the tape head is positioned over a single square.

**Initially, the machine is in state 0, the head is over square 1, the
input string x is in squares 1 to n and $C(x)$ (the input certificate
of x) is in squares 0, -1, ..., $-P(n)$.**

At step $p(n)$ the machine is in state q_Y.

The last set of restrictions is a little trickier:

**At each step the machine moves to its next configuration (state,
symbol, head position) in accordance with the application of its
program module to its previous (state, symbol).**

To find the clauses that will do this job, consider first the following
condition: the symbol in square j of the tape cannot change during step i
of the computation if the tape head isn't positioned there at that moment.
This translates into the collection

$$\{T_{i,j}, \overline{S}_{i,j,k}, S_{i+1,j,k}\}$$

of clauses, one for each triple $(i, j, k) = $ (state, square, symbol). These
clauses express the condition in the following way: either (at time i) the
tape head is positioned over square j ($T_{i,j}$ is true) or else the head is not
positioned there, in which case either symbol k is not in the jth square
before the step or else symbol k is (still) in the jth square after the step is
executed.

It remains to express the fact that the transitions from one configuration of the machine to the next are the direct results of the operation of
the program module. The three sets of clauses that do this are

$$\{\overline{T}_{i,j}, \overline{Q}_{i,k}, \overline{S}_{i,j,l}, T_{i+1,j+INC}\}$$
$$\{\overline{T}_{i,j}, \overline{Q}_{i,k}, \overline{S}_{i,j,l}, Q_{i+1,k'}\}$$
$$\{\overline{T}_{i,j}, \overline{Q}_{i,k}, \overline{S}_{i,j,l}, S_{i+1,j,l'}\}.$$

In each case the format of the clause is this: 'either the tape head is
not positioned at square j, or the present state is not q_k or the symbol just
read is not l, but if they are, then ...' There is a clause as above for each
step $i = 0, P(n)$ of the computation, for each square $j = -P(n), P(n)$ of
the tape, for each symbol l in the alphabet, and for each possible state q_k of

the machine, a polynomial number of clauses in all. The new configuration triple (INC, k', l') is, of course, as computed by the program module.

Now we have constructed a collection of clauses with the following property. If we execute a recognizing computation on a string x and its certificate, in time at most $P(n)$, then this computation determines a set of ($True$, $False$) values for all of the variables listed above, in such a way that all of the clauses just constructed are simultaneously satisfied.

Conversely, if we have a set of values of the SAT variables that satisfy all of the clauses at once, then that set of values of the variables describes a certificate that would cause TMQ to do a computation that would recognize the string x and it also describes, in minute detail, the ensuing accepting computation that TMQ would do if it were given x and that certificate.

Hence every language in NP can be reduced to SAT. It is not difficult to check through the above construction and prove that the reduction is accomplishable in polynomial time. It follows that SAT is NP-complete. ∎

5.4 Some other NP-complete problems

Cook's theorem opened the way to the identification of a vast number of NP-complete problems. The proof that Satisfiability is NP-complete required a demonstration that every problem in NP is polynomially reducible to SAT. To prove that some other problem X is NP-complete it will be sufficient to show that SAT reduces to problem X. For if that is so then every problem in NP can be reduced to problem X by first reducing to an instance of SAT and then to an instance of X.

In other words, life after Cook's theorem is a lot easier. To prove that some problem is NP-complete we need show only that SAT reduces to it. We don't have to go all the way back to the Turing machine computations any more. Just prove that if you can solve your problem then you can solve SAT. By Cook's theorem you will then know that by solving your problem you will have solved every problem in NP.

For the honor of being 'the second NP-complete problem,' consider the following special case of SAT, called *3-satisfiability*, or *3SAT*. An instance of 3SAT consists of a number of clauses, just as in SAT, except that the clauses are permitted to contain no more than 3 literals each. The question, as in SAT, is 'Are the clauses simultaneously satisfiable by some assignment of *True*, *False* values to the variables?'

Interestingly, though, the general problem SAT is reducible to the apparently more special problem 3SAT, which will show us

Theorem 5.4.1. *3-satisfiability is NP-complete.*

Proof: Let an instance of SAT be given. We will show how to transform it, quickly, to an instance of 3SAT that is satisfiable if and only if the original SAT problem was satisfiable.

More precisely, we are going to replace clauses that contain more than 3 literals with collections of clauses that contain exactly 3 literals and that have the same satisfiability as the original. In fact, suppose our instance of SAT contains a clause

$$\{x_1, x_2, \ldots, x_k\} \quad (k \geq 4). \tag{5.4.1}$$

Then this clause will be replaced by exactly $k - 2$ new clauses, utilizing $k - 3$ new variables z_i $(i = 1, k - 3)$ that are introduced just for this purpose. The $k - 2$ new clauses are

$$\{x_1, x_2, z_1\}, \{x_3, \bar{z}_1, z_2\}, \{x_4, \bar{z}_2, z_3\}, \ldots, \{x_{k-1}, x_k, \bar{z}_{k-3}\}. \tag{5.4.2}$$

We now make the following

Claim. *If* $x_1^*, \ldots, x_k^*$ *is an assignment of truth values to the x's for which the clause (5.4.1) is true, then there exist assignments* $z_1^*, \ldots, z_{k-3}^*$ *of truth values to the z's such that all of the clauses (5.4.2) are simultaneously satisfied by (x^*, z^*), and conversely, if (x^*, z^*) is some assignment that satisfies all of (5.4.2), then x^* alone satisfies (5.4.1).*

To prove the claim, first suppose that (5.4.1) is satisfied by some assignment x^*. Then one, at least, of the k literals $x_1, \ldots, x_k$, say x_r, has the value *True*.

Then we can satisfy all $k - 2$ of the transformed clauses (5.4.2) by assigning $z_s^* := True$ for $s \leq r - 2$ and $z_s^* := False$ for $s > r - 2$. It is easy to check that each one of the $k - 2$ new clauses is satisfied.

Conversely, suppose that all of the new clauses are satisfied by some assignment of truth values to the x's and the z's. We will show that at least one of the x's must be *True*, so that the original clause will be satisfied.

Suppose, to the contrary, that all of the x's are *False*. Since, in the new clauses none of the x's are negated, the fact that the new clauses are satisfied tells us that they would remain satisfied without any of the x's. Hence the clauses

$$\{z_1\}, \{\bar{z}_1, z_2\}, \{\bar{z}_2, z_3\}, \ldots, \{\bar{z}_{k-4}, z_{k-3}\}, \{\bar{z}_{k-3}\}$$

are satisfied by the values of the z's. If we scan the list from left to right we discover, in turn, that z_1 is true, z_2 is true, ..., and finally, much to our surprise, that z_{k-3} is true, and z_{k-3} is also false, a contradiction, which establishes the truth of the claim made above.

The observation that the transformations just discussed can be carried out in polynomial time completes the proof of theorem 5.4.1. ■

We remark, in passing, that the problem '2SAT' is in P.

Our collection of NP-complete problems is growing. Now we have two, and a third is on the way. We will show next how to reduce 3SAT to a graph-coloring problem, thereby proving

Theorem 5.4.2. *The graph vertex-coloring problem is NP-complete.*

Proof: Given an instance of 3SAT, that is to say, given a collection of k clauses, involving n variables and having at most three literals per clause, we will construct, in polynomial time, a graph G with the property that its vertices can be properly colored in $n + 1$ colors if and only if the given clauses are satisfiable. We will assume that $n > 4$, the contrary case being trivial.

The graph G will have $3n + k$ vertices:

$$\{x_1, \ldots, x_n\}, \{\bar{x}_1, \ldots, \bar{x}_n\}, \{y_1, \ldots, y_n\}, \{C_1, \ldots, C_k\}$$

Now we will describe the set of edges of G. First, each vertex x_i is joined to $\bar{x}_i$ $(i = 1, n)$. Next, every vertex y_i is joined to every other vertex y_j $(j \neq i)$, to every other vertex x_j $(j \neq i)$, and to every other vertex $\bar{x}_j$ $(j \neq i)$.

Vertex x_i is connected to C_j if x_i is *not* one of the literals in clause C_j. Finally, $\bar{x}_i$ is connected to C_j if $\bar{x}_i$ is not one of the literals in C_j.

May we interrupt the proceedings to say again why we're doing all of this? You have just read the description of a certain graph G. The graph is one that can be drawn as soon as someone hands us a 3SAT problem. We described the graph by listing its vertices and then listing its edges. What does the graph do for us?

Well, suppose that we have just bought a computer program that can decide if graphs are colorable in a given number of colors. We paid $49.95 for it, and we'd like to use it. But the first problem that needs solving happens to be a 3SAT problem, not a graph-coloring problem. We aren't so easily discouraged, though. We convert the 3SAT problem into a graph that is $(n + 1)$-colorable if and only if the original 3SAT problem was satisfiable. Now we can get our money's worth by running the graph-coloring program even though what we really wanted to do was to solve a 3SAT problem.

In Fig. 5.4.1 we show the graph G of 11 vertices that corresponds to the following instance of 3SAT:

$$C_1 := \{x_1, \bar{x}_2\}; \quad C_2 := \{x_1, x_2, \bar{x}_3\}.$$

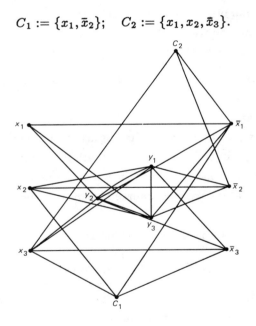

Fig. 5.4.1: The graph for a 3SAT problem

Now we claim that this graph is $n+1$ colorable if and only if the clauses are satisfiable.

Clearly G cannot be colored in fewer than n colors, because the n vertices $y_1, \ldots, y_n$ are all connected to each other and therefore they alone already require n different colors for a proper coloration. Suppose that y_i is assigned color i $(i = 1, n)$.

Do we need new colors in order to color the x_i vertices? Since vertex y_i is connected to every x vertex and every $\overline{x}$ vertex except x_i, $\overline{x}_i$, if color i is going to be used on the x's or the $\overline{x}$'s, it will have to be assigned to one of x_i, $\overline{x}_i$, but not to both, since they are connected to each other. Hence a new color, color $n + 1$, will have to be introduced in order to color the x's and $\overline{x}$'s.

Further, if we are going to color the vertices of G in only $n + 1$ colors, the only way to do it will be to assign color $n + 1$ to exactly one member of each pair $(x_i, \overline{x}_i)$, and color i to the other one, for each $i = 1, n$. That one of the pair that gets color $n + 1$ will be called the *False* vertex, the other one is the *True* vertex of the pair $(x_i, \overline{x}_i)$, for each $i = 1, n$.

It remains to color the vertices $C_1, \ldots, C_k$. The graph will be $n + 1$ colorable if and only if we can do this without using any new colors. Since each clause contains at most 3 literals, and $n > 4$, every variable C_i must be adjacent to both x_j and $\overline{x}_j$ for at least one value of j. Therefore no vertex C_i can be colored in the color $n + 1$ in a proper coloring of G, and therefore every C_i must be colored in one of the colors $1, 2, \ldots, n$.

Since C_i is connected by an edge to every vertex x_j or $\overline{x}_j$ that is not in the clause C_i, it follows that C_i cannot be colored in the same color as any x_j or $\overline{x}_j$ that is not in the clause C_i.

Hence the color that we assign to C_i must be the same as the color of some '*True*' vertex x_j or $\overline{x}_j$ that corresponds to a literal that is in clause C_i. Therefore the graph G is $n + 1$ colorable if and only if there is a '*True*' vertex for each C_i, and this means exactly that the clauses are satisfiable.

It is easy to verify that the transformation from the 3SAT problem to the graph-coloring problem can be carried out in polynomial time, and the proof is finished. ∎

By means of many, often quite ingenious, transformations of the kind that we have just done, the list of NP-complete problems has grown rapidly since the first example, and the 21 additional problems found by R. Karp. Hundreds of such problems are now known. Here are a few of the more important ones.

Maximum clique: We are given a graph G and an integer K. The question is to determine whether or not there is a set of K vertices in G each of which is adjacent, by an edge of G, to all of the others.

Edge coloring: Given a graph G and an integer K. Can we color the *edges* of G, in K colors, so that whenever two edges meet at a vertex, they will have different colors?

Let us refer to an edge coloring of this kind as a *proper* coloring of the edges of G.

A beautiful theorem of Vizing* deals with this question. If Δ denotes the largest degree of any vertex in the given graph, then Vizing's theorem asserts that the edges of G can be properly colored in either Δ or $\Delta + 1$ colors. Since it is obvious that *at least* Δ colors will be needed, this means that the edge chromatic number is in doubt by only one unit, for every graph G! Nevertheless, the decision as to whether the correct answer is Δ or $\Delta + 1$, is NP-complete.

Hamilton path: In a given graph G, is there a path that visits every vertex of G exactly once?

Target sum: Given a finite set of positive integers whose sum is S. Is there a subset whose sum is $S/2$?

The above list, together with SAT, 3SAT, Travelling Salesman and Graph Coloring, constitutes a modest sampling of the class of these seemingly intractable problems. Of course it must not·be assumed that every problem that 'sounds like' an NP-complete problem is necessarily so hard. If, for example, we ask for an Euler path instead of a Hamilton path (*i.e.*, if we want to traverse *edges* instead of *vertices*, the problem would no longer be NP-complete, and in fact, it would be in P, thanks to Theorem 1.6.1.

As another example, the fact that one can find the edge connectivity of a given graph in polynomial time (see section 3.8) is rather amazing, considering the quite difficult appearance of the problem. One of our motivations for including the network flow algorithms in this book was, indeed, to show how very sophisticated algorithms can sometimes prove that seemingly hard problems are in fact computationally tractable.

---

* V. G. Vizing, On an estimate of the chromatic class of a p-graph (Russian), *Diskret. Analiz.* **3** (1964), 25-30.

Exercises for section 5.4

1. Is the claim that we made and proved above (just after (5.4.2)) identical with the statement that the clause (5.4.1) is satisfiable if and only if the clauses (5.4.2) are simultaneously satisfiable? Discuss.

2. Is the claim that we made and proved above (just after (5.4.2)) identical with the statement that the Boolean expression (5.4.1) is equal to the product of the Boolean expressions (5.4.2) in the sense that their truth values are identical on every set of inputs? Discuss.

3. Let it be desired to find out if a given graph G, of V vertices, can be vertex colored in K colors. If we transform the problem into an instance of 3SAT, exactly how many clauses will there be?

5.5 Half a loaf ...

If we simply *have* to solve an NP-complete problem, then we are faced with a very long computation. Is there anything that can be done to lighten the load? In a number of cases various kinds of probabilistic and approximate algorithms have been developed, some very ingenious, and these may often be quite serviceable, as we have already seen in the case of primality testing. Here are some of the categories of 'near' solutions that have been developed.

Type I: *'Almost surely...'*

Suppose we have an NP-complete problem that asks if there is a certain kind of substructure embedded inside a given structure. Then we may be able to develop an algorithm with the following properties:

(a) It always runs in polynomial time

(b) When it finds a solution then that solution is always a correct one

(c) It doesn't always find a solution, but it 'almost always' does, in the sense that the ratio of successes to total cases approaches unity as the size of the input string grows large.

An example of such an algorithm is one that will find a Hamilton path in almost all graphs, failing to do so sometimes, but not often, and running always in polynomial time. We will describe such an algorithm below.

Type II: *'Usually fast...'*

In this category of quasi-solution are algorithms in which the uncertainty lies not in whether a solution will be found, but in how long it will take to find one. An algorithm of this kind will

(a) always find a solution, and the solution will always be correct, and

(b) operate in an *average* time that is better than exponential, although occasionally it may require exponential time. The averaging is over all input strings of given size.

An example of this sort is an algorithm that will surely find a maximum independent set in a graph, will on the average require 'only' $O(n^{c \log n})$ time to do so, but will occasionally (*i.e.*, for some graphs) require nearly 2^n time to get an answer. We will outline such an algorithm below, in section 5.6. Note that $O(n^{c \log n})$ is not a polynomial-time estimate, but it's an improvement over 2^n.

Type III: *'Approximately the right answer...'*

In this kind of an algorithm we don't even get the right answer, but it's close. Since this means giving up quite a bit, people like these algorithms to be very fast. Of course we are going to drop our insistence that the questions be posed as decision problems, and instead they will be asked as optimization problems: find the shortest tour through these cities, or, find the size of the maximum clique in this graph, or, find a coloring of this graph in the fewest possible colors, etc.

In response, these approximation algorithms will

(a) run in polynomial time

(b) always produce some output

(c) provide a guarantee that the output will not deviate from the optimal solution by more than such-and-such.

An example of this type is the approximate algorithm for the travelling salesman problem that is given below, in section 5.8. It quickly yields a tour of the cities that is guaranteed to be at most twice as long as the shortest possible tour.

Now let's look at examples of each of these kinds of approximation algorithms.

An example of an algorithm of Type I is due to Angluin and Valiant. It tries to find a Hamilton path (or circuit) in a graph G. It doesn't always find such a path, but in theorem 5.5.1 below we will see that it usually does, at least if the graph G is from a class of graphs that are likely to have Hamilton paths at all.

Input to the algorithm are the graph G and two distinguished vertices s, t. It looks for a Hamilton path between the vertices s, t (if $s = t$ on input, then we are looking for a Hamilton circuit in G).

The procedure maintains a partially constructed Hamilton path P, from s to some vertex ndp, and it attempts to extend P by adjoining an edge to a new, previously unvisited vertex. In the process of doing so it will delete from the graph G, from time to time, an edge, so we will also maintain a variable graph G', that is initially set to G, but which is acted upon by the program.

To do its job, the algorithm chooses at random an edge (ndp, v) that is incident with the current endpoint of the partial path P, and it deletes the edge (ndp, v) from the graph G', so it will never be chosen again. If v is a vertex that is not on the path P, then the path is extended by adjoining the new edge (ndp, v).

So much is fairly clear. However, if the new vertex v is already on the path P, then we short circuit the path by deleting an edge from it and drawing in a new edge, as is shown below in the formal statement of the algorithm, and in Fig. 5.5.1. In that case the path does not get longer, but it changes so that it now has enhanced chances of ultimate completion.

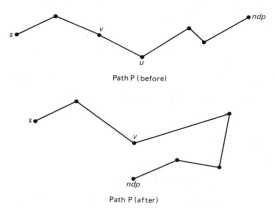

Path P (before)

Path P (after)

Fig. 5.5.1: The short circuit

Here is a formal statement of the algorithm of Angluin and Valiant for finding a Hamilton path or circuit in an undirected graph G.

procedure *uhc* (*G*:graph;*s t*:vertex);
{finds a Hamilton path (if $s \neq t$) or a Hamilton
    circuit (if $s = t$) *P* in an undirected graph *G*
      and returns '*success*' or fails, and returns '*failure*'}
G' := *G*; *ndp* := *s*; *P* := empty path;
repeat
if *ndp* is an isolated point of *G'*
    then return '*failure*'
    else
        choose uniformly at random an edge (*ndp*, *v*) from
            among the edges of *G'* that are incident with *ndp*
            and delete that edge from *G'*;
      if $v \neq t$ and $v \notin P$
        then adjoin the edge (*ndp*, *v*) to *P*; *ndp* := *v*
        else
          if $v \neq t$ and $v \in P$
            then
              {this is the short-circuit of Fig. 5.5.1}
                u:= neighbor of *v* in *P* that is closer to *ndp*;
                delete edge (*u*, *v*) from *P*;
                adjoin (*ndp*, *v*) to *P*;
                ndp := *u*
              end; {**then**}
        end{*else*}
until *P* contains every vertex of *G* (except *t*, if
      $s \neq t$) and edge (*ndp*, *t*) is in *G* but not in *G'*;
adjoin edge (*ndp*, *t*) to *P* and return '*success*'
end.{uhc}

As stated above, the algorithm makes only a very modest claim: either it succeeds or it fails! Of course what makes it valuable is the accompanying theorem, which asserts that in fact the procedure almost always succeeds, provided that the graph *G* has a good chance of having a Hamilton path or circuit.

What kind of a graph has such a 'good chance'? A great deal of research has gone into the study of how many edges a graph has to have before almost surely it must contain certain given structures. For instance, how many edges must a graph of n vertices have before we can be almost certain that it will contain a complete graph of 4 vertices?

To say that graphs have a property 'almost certainly' is to say that the ratio of the number of graphs on n vertices that have the property to the number of graphs on n vertices approaches 1 as n grows without bound.

For the Hamilton path problem, an important dividing line, or threshold, turns out to be at the level of $cn \log n$ edges. That is to say, a graph of n vertices that has $o(n \log n)$ edges has relatively little chance of being even connected, whereas a graph with $> cn \log n$ edges is almost certainly connected, and almost certainly has a Hamilton path.

We now state the theorem of Angluin and Valiant, which asserts that the algorithm above will almost surely succeed if the graph G has enough edges.

Theorem 5.5.1. *Fix a positive real number a. There exist numbers M and c such that if we choose a graph G at random from among those of n vertices and at least $cn \log n$ edges, and we choose arbitrary vertices s, t in G, then the probability that algorithm UHC returns 'success' before making a total of $Mn \log n$ attempts to extend partially constructed paths is $1 - O(n^{-a})$.*

5.6 Backtracking (I): independent sets

In this section we are going to describe an algorithm that is capable of solving some NP-complete problems fast, *on the average*, while at the same time guaranteeing that a solution will always be found, be it quickly or slowly.

The method is called *backtracking*, and it has long been a standard method in computer search problems when all else fails. It has been common to think of backtracking as a very long process, and indeed it can be. But recently it has been shown that the method can be very fast on average, and that in the graph-coloring problem, for instance, it functions in an average of *constant* time, *i.e.*, the time is independent of the number of vertices, although to be sure, the worst-case behavior is very exponential.

We will first illustrate the backtrack method in the context of a search for the largest independent set of vertices (a set of vertices no two of which are joined by an edge) in a given graph G, an NP-complete problem. In this case the average time behavior of the method is not constant, or even polynomial, but it is subexponential. The method is also easy to analyze and to describe in this case, which is why we are choosing it as our first illustration.

Hence consider a graph G, of n vertices, in which the vertices have been numbered $1, 2, \ldots, n$. We want to find, in G, the size of the largest independent set of vertices. In Fig. 5.6.1 below, the graph G has 6 vertices.

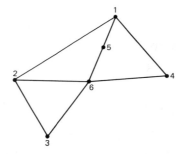

Fig. 5.6.1: Find the largest independent set

Begin by searching for an independent set S that contains vertex 1, so let $S := \{1\}$. Now attempt to enlarge S. We cannot enlarge S by adjoining vertex 2 to it, but we can add vertex 3. Our set S is now $\{1,3\}$.

Now we cannot adjoin vertex 4 (joined to 1) or vertex 5 (joined to 1) or vertex 6 (joined to 3), so we are stuck. Therefore we backtrack, by replacing the most recently added member of S by the next choice that we might have made for it. In this case, we delete vertex 3 from S, and the next choice would be vertex 6. The set S is $\{1,6\}$. Again we have a dead end.

If we backtrack again, there are no further choices with which to replace vertex 6, so we backtrack even further, and not only delete the 6 from S but also replace vertex 1 by the next possible choice for it, namely vertex 2.

To speed up the discussion, we will now show the list of all sets S that turn up from start to finish of the algorithm:

$\{1\}$, $\{13\}$, $\{16\}$, $\{2\}$, $\{24\}$, $\{245\}$, $\{25\}$, $\{3\}$, $\{34\}$, $\{345\}$, $\{35\}$, $\{4\}$, $\{45\}$, $\{5\}$, $\{6\}$

A convenient way to represent the search process is by means of the *backtrack search tree T*. This is a tree whose vertices are arranged on levels $L := 0, 1, \ldots, n$ for a graph of n vertices. Each vertex of T corresponds to an independent set of vertices in G. Two vertices of T, corresponding to independent sets S', S'' of vertices of G, are joined by an edge in T if $S' \subset S''$, and $S'' - S'$ consists of a single element: the highest-numbered vertex in S''. On level L we find a vertex S of T for every independent set of exactly L vertices of G. Level 0 consists of a single root vertex, corresponding to the empty set of vertices of G.

The complete backtrack search tree for the problem of finding a maximum independent set in the graph G of Fig. 5.6.1 is shown in Fig. 5.6.2 below.

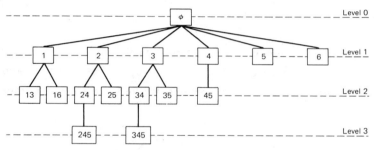

Fig. 5.6.2: The backtrack search tree

The backtrack algorithm amounts just to visiting every vertex of the search tree T, without actually having to write down the tree explicitly, in advance.

Observe that the list of sets S above, or equivalently, the list of nodes of the tree T, consists of exactly *every independent set in the graph G*. A reasonable measure of the complexity of the searching job, therefore, is the number of independent sets that G has. In the example above, the graph G had 19 independent sets of vertices, including the empty set.

The question of the complexity of backtrack search is therefore the same as the question of determining the number of independent sets of G.

Some graphs G have an enormous number of independent sets. The graph $\overline{K}_n$ of n vertices and no edges whatever has exactly 2^n independent sets of vertices. The backtrack tree will have 2^n nodes, and the search will be a long one indeed.

The complete graph K_n of n vertices and every possible edge, $n(n-1)/2$ in all, has just $n+1$ independent sets of vertices.

Any other graph G of n vertices will have a number of independent sets that lies between these two extremes of $n+1$ and 2^n. Sometimes backtracking will take an exponentially long time, and sometimes it will be fairly quick. Now the question is, *on the average* how fast is the backtrack method for this problem?

What we are asking for is the average number of independent sets that a graph of n vertices has. Let $I(G)$ denote the number of independent sets in the graph G, and let $H(G, S)$ be 1 if a set S of vertices is independent in G and 0 otherwise. Then obviously

$$I(G) = \sum_{S \subseteq [n]} H(G, S) \tag{5.6.1}$$

in which the sum extends over all 2^n subsets S of $[n] = \{1, \ldots, n\}$.

If we sum both sides of (5.6.1) over all graphs G of n vertices and divide by the number $2^{n(n-1)/2}$ of such graphs, we find the average number I_n of independent sets in the form

$$I_n = 2^{-n(n-1)/2} \sum_{|V(G)|=n} \left\{ \sum_{S \subseteq [n]} H(G, S) \right\}. \tag{5.6.2}$$

Now interchange the order of the two summation signs on the right to obtain

$$I_n = 2^{-n(n-1)/2} \sum_{S \subseteq [n]} \left\{ \sum_{|V(G)|=n} H(G, S) \right\}. \tag{5.6.3}$$

Let's contemplate the inner sum on the right side. In that sum the set S is fixed and we are summing over all graphs G. Hence the sum is equal to the number of graphs G of n vertices that have the particular set S as an independent set.

If S is a given set of k vertices, how many graphs G have S for an independent set? That is, how many G have no edges joining a pair of vertices in S? There are $k(k-1)/2$ edges that G is *not* allowed to have. That leaves $n(n-1)/2 - k(k-1)/2$ edges that G *might* have.

Therefore there are exactly $2^{n(n-1)/2 - k(k-1)/2}$ graphs of n vertices that have the given set S of k vertices as an independent set.

If we substitute the above value for the inner sum in (5.6.3), and use the fact that there are exactly $\binom{n}{k}$ sets S of k vertices chosen from n, then the whole expression (5.6.3) simplifies to

$$I_n = \sum_{k=0}^{n} \binom{n}{k} 2^{-k(k-1)/2}. \tag{5.6.4}$$

Hence, in (5.6.4) we have an exact formula for the average number of independent sets in a graph of n vertices. A short table of values of I_n is shown below, in Table 5.6.1, along with values of 2^n, for comparison. Clearly the average number of independent sets in a graph is a lot smaller than the maximum number that graphs of that size might have.

n	I_n	2^n
2	3.5	4
3	5.6	8
4	8.5	16
5	12.3	32
10	52.0	1024
15	149.8	32768
20	350.6	1048576
30	1342.5	1073741824
40	3862.9	1099511627776

Table 5.6.1: Independent sets and all sets

In the exercises it will be seen that the rate of growth of I_n as n grows large is $O(n^{\log n})$. Hence the average amount of labor in a backtrack search for the largest independent set in a graph grows subexponentially, although faster than polynomially. It is some indication of how hard this problem is that even on the average the amount of labor needed is not of polynomial growth.

Exercises for section 5.6

1. What is the average number of independent sets of size k that are in graphs of V vertices and E edges?

2. Let t_k denote the kth term in the sum (5.6.4).

    (a) Show that $t_k/t_{k-1} = (n - k + 1)/(k2^{k-1})$.

    (b) Show that t_k/t_{k-1} is > 1 when k is small, then is < 1 after k passes a certain critical value k_0. Hence show that the terms in the sum (5.6.4) increase in size until $k = k_0$ and then decrease.

3. Now we will estimate the size of k_0 in the previous problem.

    (a) Show that $t_k/t_{k-1} < 1$ when $k = \lfloor \log_2 n \rfloor$ and $t_k/t_{k-1} > 1$ when $k = \lfloor \log_2 n - \log_2 \log_2 n \rfloor$. Hence the index k_0 of the largest term in (5.6.4) satisfies

$$\log_2 n - \log_2 \log_2 n \le k_0 \le \log_2 n$$

    (b) The entire sum in (5.6.4) is at most $n + 1$ times as large as its largest single term. Use Stirling's formula (1.1.10) and 3(a) above, to show that the k_0^{th} term is $O((n+\epsilon)^{\log n})$ and therefore the same is true of the whole sum, *i.e.*, of I_n.

5.7 Backtracking (II): graph coloring

It has recently been discovered that in another NP-complete problem, that of graph coloring, the average amount of labor in a backtrack search is $O(1)$ (bounded) as n, the number of vertices of the graph, grows without bound. More precisely, for fixed K, if we ask 'Is the graph G, of V vertices, properly vertex-colorable in K colors?', then the average labor in a backtrack search for the answer is bounded. Hence not only is the average of polynomial growth, but the polynomial is of degree 0 (in V).

To be even more specific, consider the case of 3 colors. It is already NP-complete to ask if the vertices of a given graph can be colored in 3 colors. Nevertheless, *the average number of nodes in the backtrack search tree for this problem is about 197*, averaged over graphs of all sizes. This means that if we input a random graph of 1,000,000 vertices, and ask if it is 3-colorable, then we can expect an answer (probably 'no') after only about 197 steps of computation.

To prove this we will need some preliminary lemmas.

Lemma 5.7.1. *Let $s_1, \ldots, s_K$ be nonnegative numbers whose sum is L. Then the sum of their squares is at least L^2/K.*

Proof: We have

$$0 \le \sum_{i=1}^{K}(s_i - L/K)^2$$

$$= \sum_{i=1}^{K}(s_i^2 - 2Ls_i/K + L^2/K^2)$$

$$= \sum_{i=1}^{K} s_i^2 - 2L^2/K + L^2/K$$

$$= \sum_{i=1}^{K} s_i^2 - L^2/K. \quad \blacksquare$$

The next lemma deals with a kind of inside-out chromatic polynomial question. Instead of asking 'How many proper colorings can a given graph have?,' we ask 'How many graphs can have a given proper coloring?'

Lemma 5.7.2. *Let C be one of the K^L possible ways to color in K colors a set of L abstract vertices $1, 2, \ldots, L$. Then the number of graphs G whose vertex set is that set of L colored vertices and for which C is a proper coloring of G is at most $2^{L^2(1-1/K)/2}$.*

Proof: In the coloring C , suppose s_1 vertices get color $1, \ldots, s_K$ get color K, where of course $s_1 + \cdots + s_K = L$. If a graph G is to admit C as a proper vertex coloring then its edges can be drawn only between vertices of different colors. The number of edges that G might have is therefore

$$s_1 s_2 + s_1 s_3 + \cdots + s_1 s_K + s_2 s_3 + \cdots + s_2 s_K + \cdots + s_{K-1} s_K$$

for which we have the following estimate:

$$\sum_{1 \le i < j \le K} s_i s_j = \frac{1}{2} \sum_{i \ne j} s_i s_j$$

$$= \frac{1}{2} \left\{ \sum_{i,j=1}^{K} s_i s_j - \sum_{i=1}^{K} s_i^2 \right\}$$

$$= \frac{1}{2} (\sum s_i)^2 - \frac{1}{2} \sum s_i^2 \qquad (5.7.1)$$

$$\le L^2/2 - \frac{1}{2}(L^2/K) \quad \text{(by lemma 5.7.1)}$$

$$= L^2(1 - 1/K)/2.$$

The number of possible graphs G is therefore at most $2^{L^2(1-1/K)/2}$. $\blacksquare$

Lemma 5.7.3. *The total number of proper colorings in* K *colors of all graphs of* L *vertices is at most*

$$K^L 2^{L^2(1-1/K)/2}.$$

Proof: We are counting the pairs (G, C) where the graph G has L vertices and C is a proper K-coloring of G. If we keep C fixed and sum on G, then by lemma 5.7.2 the sum is at most $2^{L^2(1-1/K)/2}$. Since there are K^L such C's, the proof is finished. ■

Now let's think about a backtrack search for a K-coloring of a graph. Begin by using color 1 on vertex 1. Then use color 1 on vertex 2 unless $(1,2)$ is an edge, in which case use color 2. As the coloring progresses through vertices $1, 2, \ldots, L$ we color each new vertex with the lowest available color number that does not cause a conflict with some vertex that has previously been colored.

At some stage we may reach a dead end: out of colors, but not out of vertices to color. In the graph of Fig. 5.7.1 if we try to 2-color the vertices we can color vertex 1 in color 1, vertex 2 in color 2, vertex 3 in color 1 and then we'd be stuck because neither color would work on vertex 4.

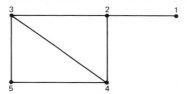

Fig. 5.7.1 Color this graph

When a dead end is reached, back up to the most recently colored vertex for which other color choices are available, replace its color with the next available choice, and try again to push forward to the next vertex.

The (futile) attempt to color the graph in Fig. 5.7.1 with 2 colors by the backtrack method can be portrayed by the *backtrack search tree* in Fig. 5.7.2.

The search is thought of as beginning at 'Root.' The label at each node of the tree describes the colors of the vertices that have so far been colored. Thus '212' means that vertices 1, 2, 3 have been colored, respectively, in colors 2, 1, 2.

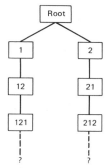

Fig. 5.7.2: A frustrated search tree

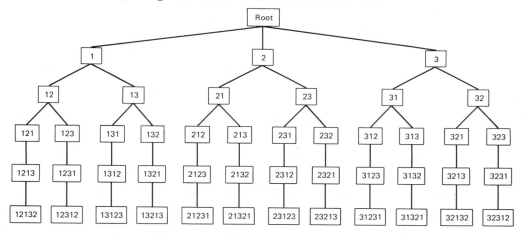

Fig. 5.7.3: A happy search tree

If instead we use 3 colors on the graph of Fig. 5.7.1 then we get a successful coloring; in fact we get 12 of them, as is shown in Fig. 5.7.3.

Let's concentrate on a particular *level* of the search tree. Level 2, for instance, consists of the nodes of the search tree that are at a distance of 2 from 'Root.' In Fig. 5.7.3, level 2 contains 6 nodes, corresponding to the partial colorings 12, 13, 21, 23, 31, 32 of the graph. When the coloring reaches vertex 2 it has seen only the portion of the graph G that is induced by vertices 1 and 2.

Generally, a node at level L of the backtrack search tree corresponds to a proper coloring in K colors of the subgraph of G that is induced by vertices $1, 2, \ldots, L$.

Let $H_L(G)$ denote that subgraph. Then we see the truth of

Lemma 5.7.4. *The number of nodes at level L of the backtrack search tree*

219

for coloring a graph G in K colors is equal to the number of proper colorings of $H_L(G)$ in K colors, i.e., to $P(K, H_L(G))$, where P is the chromatic polynomial. ∎

We are now ready for the main question of this section: what is the average number of nodes in a backtrack search tree for K-coloring graphs of n vertices? This is

$$A(n, K) = \frac{1}{no.\ of\ graphs} \sum_{graphs\ G_n} \{no.\ of\ nodes\ in\ tree\ for\ G\}$$

$$= 2^{-\binom{n}{2}} \sum_{G_n} \{\sum_{L=0}^{n} \{no.\ of\ nodes\ at\ level\ L\}\}$$

$$= 2^{-\binom{n}{2}} \sum_{G_n} \sum_{L=0}^{n} P(K, H_L(G)) \quad \text{(by lemma 5.7.4)}$$

$$= 2^{-\binom{n}{2}} \sum_{L=0}^{n} \{\sum_{G_n} P(K, H_L(G))\}.$$

Fix some value of L and consider the inner sum. As G runs over all graphs of n vertices, $H_L(G)$ selects the subgraph of G that is induced by vertices $1, 2, \ldots, L$. Now lots of graphs G of n vertices have the same $H_L(G)$ sitting at vertices $1, 2, \ldots, L$. In fact, exactly $2^{\binom{n}{2}-\binom{L}{2}}$ different graphs G of n vertices all have the same graph H of L vertices in residence at vertices $1, 2, \ldots, L$ (see exercise 15 of section 1.6). Hence (5.7.2) gives

$$A(n, K) = 2^{-\binom{n}{2}} \sum_{L=0}^{n} 2^{\binom{n}{2}-\binom{L}{2}} \left\{ \sum_{H_L} P(K, H) \right\}$$

$$= \sum_{L=0}^{n} 2^{-\binom{L}{2}} \left\{ \sum_{H_L} P(K, H) \right\}.$$

The inner sum is exactly the number that is counted by lemma 5.7.3, and so

$$A(n, K) \le \sum_{L=0}^{n} 2^{-\binom{L}{2}} K^L 2^{L^2(1-1/K)/2}$$

$$\le \sum_{L=0}^{\infty} K^L 2^{L/2} 2^{-L^2/2K}.$$

The infinite series actually converges! Hence $A(n, K)$ is bounded, for all n. This proves

Theorem 5.7.1. *Let $A(n, K)$ denote the average number of nodes in the backtrack search trees for K-coloring the vertices of all graphs of n vertices. Then there is a constant $h = h(K)$, that depends on the number of colors, K, but not on n, such that $A(n, K) \leq h(K)$ for all n.* ■

5.8 Approximate algorithms for hard problems

Finally we come to Type III of the three kinds of 'half-a-loaf-is-better-than-none' algorithms that were described in section 5.5. In these algorithms we don't find the exact solution of the problem, only an approximate one. As consolation we have an algorithm that runs in polynomial time as well as a performance guarantee to the effect that while the answer is approximate, it can deviate by no more than such-and-such from the exact answer.

An elegant example of such a situation is in the Travelling Salesman Problem, which we will now express as an optimization problem rather than as a decision problem.

We are given n points ('cities') in the plane, as well as the distances between every pair of them, and we are asked to find a round-trip tour of all of these cities that has minimum length. We will assume, throughout the following discussion that the distances satisfy the triangle inequality. This restricted version of the TSP is often called the 'Euclidean' Travelling Salesman Problem.

The algorithm that we will discuss for this problem has the properties

(a) it runs in polynomial time and

(b) the round-trip tour that it finds will never be more than twice as long as the shortest possible tour.

The first step in carrying out the algorithm is to find a minimum spanning tree (MST) for the n given cities. A MST is a tree whose nodes are the cities in question, and which, among all possible trees on that vertex set, has minimum possible length.

It may seem that finding a MST is just as hard as solving the TSP, but NIN (No, It's Not). The MST problem is one of those all-too-rare computational situations in which it pays to be greedy.

Generally speaking, in a greedy algorithm,

(i) we are trying to construct some optimal structure by adding one piece at a time, and

(ii) at each step we make the decision about which piece will be added next by choosing, among all available pieces, the single one that will carry us as far as possible in the desirable direction (be greedy!).

The reason that greedy algorithms usually are not the best possible ones is that it may be better not to take the single best piece at each step, but to take some other piece, in the hope that then at a later step we will be able to improve things even more. In other words, the *global* problem of finding the best structure might not be solvable by the *local* procedure of being as greedy as possible at every single step.

In the *MST* problem, though, the greedy strategy works, as we see in the following algorithm.

procedure *mst*(x:array of n points in the plane);
{constructs a spanning tree T of minimum length, on the
        vertices $\{x_1, x_2, \ldots, x_n\}$ in the plane}
let T consist of a single vertex, x_1;
  while T has fewer than n vertices **do**
    for each vertex v that is not yet in T, find the
        distance $d(v)$ from v to the nearest vertex of T;
    let v^* be the vertex of smallest $d(v)$;
    adjoin v^* to the vertex set of T;
    adjoin to T the edge from v^* to the nearest
        vertex $w \neq v^*$ of T;
  end{while}
end. {*mst*}

Proof of correctness of *mst*: Let T be the tree that is produced by running *mst*, and let $e_1, \ldots, e_{n-1}$ be its edges, listed in the same order in which algorithm *mst* produced them.

Let T' be a minimum spanning tree for x. Let e_r be the first edge of T that does not appear in T'. In the minimum tree T', edges $e_1, \ldots, e_{r-1}$ all appear, and we let S be the union of their vertex sets. In T' let f be the edge that joins the subtree on S to the subtree on the remaining vertices of x.

Suppose f is shorter than e_r. Then f was one of the edges that was available to algorithm *mst* at the instant that it chose e_r, and since e_r was the shortest available edge at that moment, we have a contradiction.

Suppose f is longer than e_r. Then T' would not be minimal because the tree that we would obtain by exchanging f for e_r in T' (why is it still a tree if we do that exchange?) would be shorter, contradicting the minimality of T'.

Hence f and e_r have the same length. In T' exchange f for e_r. Then T' is still a tree, and is still a minimum spanning tree.

The index of the first edge of T that does not appear in T' is now at least $r + 1$, one unit larger than it was before. The process of replacing edges of T that do not appear in T' without affecting the minimality of T can be repeated until every edge of T appears in T', *i.e.*, until $T = T'$. Hence T was a minimum spanning tree. ∎

That finishes one step of the process that leads to a polynomial-time travelling salesman algorithm that finds a tour of at most twice the minimum length.

The next step involves finding an Euler circuit. Way back in theorem 1.6.1 we learned that a connected graph has an Eulerian circuit if and only if every vertex has even degree. Recall that the proof was recursive in nature, and immediately implies a linear-time algorithm for finding Eulerian circuits recursively. We also noted that the proof remains valid even if we are dealing with a *multigraph*, that is, with a graph in which several edges are permitted between single pairs of vertices. We will in fact need that extra flexibility for the purpose at hand.

Now we have the ingredients for the quick near-optimal travelling salesman tour.

Theorem 5.8.1. *There is an algorithm that operates in polynomial time and which will return a travelling salesman tour whose length is at most twice the length of a minimum tour.*

Here is the algorithm. Given the n cities in the plane:

(1) Find a minimum spanning tree T for the cities.

(2) Double each edge of the tree, thereby obtaining a 'multitree' $T^{(2)}$ in which between each pair of vertices there are 0 or 2 edges.

(3) Since every vertex of the doubled tree has even degree, there is an

Eulerian tour W of the edges of $T^{(2)}$; find one, as in the proof of theorem 1.6.1.

(4) Now we construct the output tour of the cities. Begin at some city and follow the walk W. However, having arrived at some vertex v, go from v directly (*via* a straight line) to the next vertex of the walk W that you haven't visited yet. This means that you will often *short-circuit* portions of the walk W by going directly from some vertex to another one that is several edges 'down the road.'

The tour Z' that results from (4) above is indeed a tour of all of the cities in which each city is visited once and only once. We claim that its length is at most twice optimal.

Let Z be an optimum tour, and let e be some edge of Z. Then $Z - e$ is a path that visits all of the cities. Since a path is a tree, $Z - e$ is a spanning tree of the cities, hence $Z - e$ is at least as long as T is, and so Z is surely at least as long as T is.

Next consider the length of the tour Z'. A step of Z' that walks along an edge of the walk W has length equal to the length of that edge of W. A step of Z' that short-circuits several edges of W has length at most equal to the sum of the lengths of the edges of W that were short-circuited. If we sum these inequalities over all steps of Z' we find that the length of Z' is at most equal to the length of W, which is in turn twice the length of the tree T.

If we put all of this together we find

$$length(Z) > length(Z - e)$$
$$\geq length(T)$$
$$= (1/2)length(W)$$
$$\geq (1/2)length(Z')$$

as claimed (!) ∎

More recently it has been proved (Christofides, 1976) that in polynomial time we can find a TSP tour whose total length is at most 3/2 as long as the minimum tour. The algorithm makes use of Edmonds' algorithm for maximum matching in a general graph (see the reference at the end of Chapter 3). It will be interesting to see if the factor 3/2 can be further refined.

Polynomial-time algorithms are known for other NP-complete problems that guarantee that the answer obtained will not exceed, by more than a constant factor, the optimum answer. In some cases the guarantees apply to the *difference* between the answer that the algorithm gives and the best one. See the references below for more information.

Exercises for section 5.8

1. Consider the following algorithm:

```
procedure mst2(x:array of n points in the plane);
{allegedly finds a tree of minimum total length that
     visits every one of the given points}
  if n = 1
    then T = {x₁}
    else
       T := mst2(n − 1,x−xₙ);
       let u be the vertex of T that is nearest to xₙ;
       mst2:=T plus vertex xₙ plus edge (xₙ, u)
  end.{mst2}
```

Is this algorithm a correct recursive formulation of the minimum spanning tree greedy algorithm? If so then prove it, and if not then give an example of a set of points where *mst2* gets the wrong answer.

Bibliography

Before we list some books and journal articles it should be mentioned that research in the area of NP-completeness is moving rapidly, and the state of the art is changing all the time. Readers who would like updates on the subject are referred to a series of articles that have appeared in each issue of the *Journal of Algorithms* in recent years. These are called 'NP-completeness: An ongoing guide.' They are written by David S. Johnson, and each of them is a thorough survey of recent progress in one particular area of NP-completeness research. They are written as updates of reference [1] below.

Journals that contain a good deal of research on the areas of this chapter include the *Journal of Algorithms*, the *Journal of the Association for Computing Machinery*, the *SIAM Journal of Computing*, *Information Processing Letters*, *SIAM Journal of Algebraic and Discrete Methods*.

The most complete reference on NP-completeness is

M. Garey and D. S. Johnson, *Computers and Intractibility; A guide to the Theory of NP-completeness*, W. H. Freeman and Co., San Francisco, 1979.

The above is highly recommended. It is readable, careful, and complete.

The earliest ideas on the computational intractability of certain problems go back to

Alan Turing, On computable numbers, with an application to the *Entscheidungsproblem*, *Proc. London Math. Soc.*, Ser. 2, **42** (1936), 230-265.

Cook's theorem, which originated the subject of NP-completeness, is in

S. A. Cook, The complexity of theorem proving procedures, *Proc. Third Annual ACM Symposium on the Theory of Computing*, ACM, New York, 1971, 151-158.

After Cook's work was done, a large number of NP-complete problems were found by

Richard M. Karp, Reducibility among combinatorial problems, in R. E. Miller and J. W. Thatcher, eds., *Complexity of Computer Computations*, Plenum, New York, 1972, 85-103.

The above paper is recommended both for its content and its clarity of presentation.

The approximate algorithm for the travelling salesman problem is in

D. J. Rosenkrantz, R. E. Stearns and P. M. Lewis, An analysis of several heuristics for the travelling salesman problem, *SIAM J. Comp.*, **6**, 1977, 563-581.

Another approximate algorithm for the Euclidean TSP, which guarantees that the solution found is no more than 3/2 as long as the optimum tour, was found by

N. Christofides, Worst case analysis of a new heuristic for the travelling salesman problem, Technical Report, Graduate School of Industrial Administration, Carnegie-Mellon University, Pittsburgh, 1976.

The minimum spanning tree algorithm is due to

R. C. Prim, Shortest connection networks and some generalizations, *Bell System Tech. J.*, **36** (1957) 1389-1401.

The probabilistic algorithm for the Hamilton path problem can be found in

D. Angluin and L. G. Valiant, Fast probabilistic algorithms for Hamilton circuits and matchings, *Proc. Ninth Annual ACM Symposium on the Theory of Computing*, ACM, New York, 1977.

The result that the graph-coloring problem can be done in constant average time is due to

H. S. Wilf, Backtrack: An $O(1)$ average time algorithm for the graph coloring problem, *Information Processing Letters* **18** (1984), 119-122.

Further refinements of the above result can be found in

E. Bender and H. S. Wilf, A theoretical analysis of backtracking in the graph-coloring problem, *Journal of Algorithms* **6** (1985), 275-282.

If you enjoyed the average numbers of independent sets and average complexity of backtrack, you might enjoy the subject of random graphs. An excellent source for this subject, in which the author shows both a firm grip on the subject and a delicious sense of humor, is

Edgar M. Palmer, *Graphical Evolution, An introduction to the theory of random graphs*, Wiley-Interscience, New York, 1985.

A review article on ideas of averaging in analysis of algorithms is

Herbert S. Wilf, Some examples of combinatorial averaging, *American Mathematical Monthly* **92** (1985), 250-260.

Index

adjacent 40
Adleman, L. 149, 164, 165, 176
Aho, A. V. 103
Angluin, D. 208-211, 227
Appel, K. 69
average complexity 57, 211*ff.*

backtracking 211*ff.*
Bender, E. 227
Bentley, J. 54
Berger, R. 3
big oh 9
binary system 19
bin-packing 178
binomial theorem 37
bipartite graph 44, 182
binomial coefficients 35
—, growth of 38
blocking flow 124
Burnside's lemma 46

cardinality 35
canonical factorization 138
capacity of a cut 115
Carmichael numbers 158
certificate 171, 182, 193
Cherkassky, B. V. 135
Chinese remainder theorem 154
chromatic number 44
chromatic polynomial 73
Cohen, H. 176
coloring graphs 43
complement of a graph 44
complexity 1
—, worst-case 4
connected 41
Cook, S. 187, 194-201, 226
Cook's theorem 195*ff.*
Cooley, J. M. 103
Coppersmith, D. 99
cryptography 165
Cristofides, N. 224, 227
cut in a network 115
—, capacity of 115
cycle 41
cyclic group 152

decimal system 19
decision problem 181

degree of a vertex 40
deterministic 193
Diffie, W. 176
digraph 105
Dinic, E. 108, 134
divide 137
Dixon, J. D. 170, 175, 177
domino problem 3

'easy' computation 1
edge coloring 206
edge connectivity 132
Edmonds, J. 107, 134, 224
Enslein, K. 103
Euclidean algorithm 140, 168
—, complexity 142
—, extended 144*ff.*
Euler totient function 138, 157
Eulerian circuit 41
Even, S. 135
exponential growth 13

factor base 169
Fermat's theorem 152, 159
FFT, complexity of 93
—, applications of 95 *ff.*
Fibonacci numbers 30, 76, 144
flow 106
—, value of 106
—, augmentation 109
—, blocking 124
flow augmenting path 109
Ford-Fulkerson algorithm 108*ff.*
Ford, L. 107*ff.*
four-color theorem 68
Fourier transform 83*ff.*
—, discrete 83
—, inverse 96
Fulkerson, D. E. 107*ff.*

Galil, Z. 135
Gardner, M. 2
Garey, M. 188
geometric series 23
Gomory, R. E. 136
graphs 40*ff.*
—, coloring of 43, 183, 216*ff.*
—, connected 41
—, complement of 44

229